T5-CSA-923

TINPLATE TOYS

FROM SCHUCO, BING, & OTHER COMPANIES

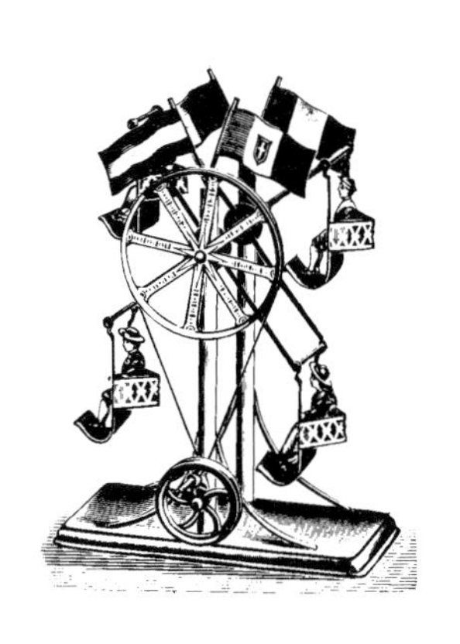

TINPLATE TOYS

FROM SCHUCO, BING, & OTHER COMPANIES

Jürgen Franzke, Editor

77 Lower Valley Road, Atglen, PA 19310

Copyright ©1995 by Schiffer Publishing, Ltd.

All rights reserved. No part of this work may be reproduced or used in any forms or by any means--graphic, electronic, or mechanical, including photocopying or information storage and retrieval systems--without written permission from the copyright holder.

This book is meant only for personal home use and recreation. It is not intended for commercial applications or manufacturing purposes.

Photography: Marcus Keresztes, Jürgen Franzke
Translated from German by Dr. Edward Force.
Originally published as *Schuco, Bing & Co. Berühmtes Blechspielzeug aus Nürnberg Band 1* by W. Tümmels Buchdruckerei und Verlag GmbH, Nürnberg.
Translation © Copyright 1995 by Schiffer Publishing, Ltd.

Printed in the United States of America

ISBN: 0-88740-863-X

Library of Congress Cataloging-in-Publication Data

Schuco, Bing & Co. berühmtes Blechspielzeug aus Nürnberg. English.
Tinplate toys : from Schuco, Bing & other companies / Jürgen Franzke, editor.
p. cm. -- (A Schiffer book for collectors)
Includes bibliographical references and index.
ISBN 0-88740-863-X (paper)
1. Tin toys--Collectors and collecting--Germany--Nuremberg--Catalogs. 2. Tin toys--Germany--Nuremberg--History--20th century--Catalogs. I. Franzke, Jürgen. II. Title. III. Series.
NK8454.2.G32N878 1995
688.7'28'0943324--dc20 95-35254
CIP

Published by Schiffer Publishing, Ltd.
77 Lower Valley Road
Atglen, PA 19310
Please write for a free catalog.
This book may be purchased from the publisher.
Please include $2.95 postage.
Try your bookstore first.

We are interested in hearing from authors
with book ideas on related subjects.

CONTENTS

INTRODUCTION

Nuremberg is a city that has worldwide renown for its toy products. In the 19th century the region grew as the center of the toy industry and has remained so to this day. The international toy fair, with its more than 2400 exhibitors, is the largest fair of its kind in the world. The Nuremberg and Fürth toy firms--Arnold, Fleischmann, Lehmann, Playmobil, Carrera and Trix--have international reputations. Along with pencils, it is surely toys that have carried the Nuremberg name into all the world.

The era of the tinplate toy began in the mid-19th century, and in the decades that followed, those toys were to dominate the playroom. The Nuremberg firm Bing already had more than 2700 employees by the turn of the century and proudly called itself "the greatest toy firm in the world." The name "Schuco" may be even more famous from a present-day point of view. In the fifties and Sixties, Schuco tinplate cars were exported all over the world, and children from America to Japan recognized this "magical" name.

There were more than fifty firms in and around Nuremberg at one time, and very little is known about most of them. This illustrated volume attempts to use striking photographs to give an initial insight into the golden age of tinplate toys. Twenty-seven firms and their offerings are presented--particularly automobiles, figures and railroads--with the two great names in the spotlight: Schuco and Bing.

ACKNOWLEDGEMENTS

The author was encouraged while preparing this book by the enthusiasm of the *Sammlergemeinschaft Blechspielzeug* of Nuremberg. It deserves my special thanks here and now; especially Mr. Wolfgang Dittrich, the tireless initiator, as well as the collectors: Fritz Ferschl, Ralf Regnat, Manfred Seeberger and Klaus Seufert.

I would also like to thank Mr. Horst E. Schmidmer and his family who made their Bing collection available to me. Helmut Kellermann provided detailed information about sheet metal production; the Toy Museum of Nuremberg, with its directoress, Dr. Lydia Bayer, and Frau Dr. Faber, Frau Dr. Kubach-Reutter and Mr. Weiss, also made museum pieces available. As did Ms. Kupiak of the Brandstätter firm. Marcus Keresztes took the excellent photographs and did much to support the project. Anne Kraft and Martha Schmitt of the Forte Design Group were responsible for the setup. Reiner Niebauer of Tümmels-Verlag took the financial risks. Great thanks and recognition are deserved by all of them. And finally, hearty, very personal thanks to my wife Regine, who helped me surmount many obstacles in a short time.

Jürgen Franzke

Chapter 1.

FAMOUS TINPLATE TOYS FROM NUREMBERG

Jürgen Franzke

As early as the 16th century there were craftsmen, the Nuremberg *Mechanici*, working at shaping and stamping sheet metal, particularly brass. All sorts of things were made of it, including toys. The artistic locksmith Caspar Werner, who died in 1545, "made, among other beautiful works of art, a ship, 3/4 of an ell long, which ran around the table with the help of several small wheels, and in which sat a feminine person, who beat the time on a cymbal with both hands, while at the front of the ship there stood a child who moved its head and paddled with both hands, and at the stern was a Cupid with his strung bow and an arrow lying on it, with such a mechanical device that you could turn it toward any person at the table, any way you wanted, and shoot the arrow at him."[1]

To be sure, we no longer possess this original ship, but the description shows us how clever, and at the same time unrealistic, it was. A toy? In any case, it is a very early example of the artisan's art of making small things. "Nuremberg, big in small things, small in big things," a saying that is often used to describe "the Nuremberger," has been applied often to the history of Nuremberg. The example of toy production is very well suited to typify this motto. Wilhelm Heinrich Wackenroder, one of the great German Romanticists, describes Nuremberg's business zeal in one of his travel accounts: "There are scarcely any large manufacturing or other factories in Nuremberg. The wares, especially the short ones (he means the small ones; J. F.), the goods for which the present-day workers are so renowned, are made by individual artisans, who have few or no assistants... The merchants then send them in great quantities to all of Europe, and because of this supplying of a whole continent with all kinds of small items, Nuremberg has actually acquired a special status among all business and trade centers."[2]

In this letter to his parents, written on October 4, 1793, Wackenroder characterizes the Nuremberg artisans as being famous for the "little things" that they make. This refers particularly to the production of toys. Wackenroder describes the activity of a maker of wooden toys: "At a woodcarver's shop I saw wooden dolls and toys, such as are also made in Fürth, Erlangen and several other places in the Empire. They are made finely by hand. The faces, hands, feet and other delicate parts, sometimes also entire figures, are formed of a hardened brown material of flour and plaster. The figures carved almost completely of wood also are not made of one piece, but out of several that are glued together."[3] For Wackenroder's exact observation we are indebted to a very precise technical-practical description of the making of wooden toys at that time. He had also visited a "Mechanikus", and wrote: "My second visit today, to the Mechanikus Mr. David Beringer (in a small house on the Steig) turned out very fortunately. Mr. Beringer is the best, or the only, Mechanikus in Nuremberg (although those who make toys, mathematical and

physical objects and such wares for sale, as well as optical cards, etc., also call themselves Mechanici and Optici).[4]
Wackenroder, who had a thorough understanding of technical things, ranked Beringer so highly because he had made "an earthly and heavenly sphere" with high precision and the greatest *akkuratesse* (accuracy). But what is interesting in terms of our subject is the comment that the Mechanici also produced toys and physical and mathematical objects. The very inclusive lists and catalogs of the "fancy goods dealer" Georg Hieronimus Bestelmeier offer a great variety of technical toys, made by Nuremberg artisans. The illustrations were made in part from copperplates. More and more Mechanici turned to the production of mechanical toys; as the 19th century began, the following items are listed by Bestelmeier: "A pump fountain with four spring pipes, nicely painted, 1 Gulden, 30 Kreuzer; a hungry swan with a magnet 36 Kreuzer; a magic platter with six animals 1 Gulden, 24 Kreuzer; an English coach which, when it is wound up, moves automatically, costs 7 Gulden with two horses; a garden vessel in which one can watch garden seeds grow very quickly, 1 Gulden, 6 Kreuzer; a telegraph as a complete model, this machine on a nicely painted and lacquered building, 15 Gulden; a mountain-climbing cylinder, 36 Kreuzer; a squirrel running by clockwork, 12 Gulden."[5]

By the end of the 18th century, Nuremberg already was renowned as a toymaking city. Tinplate figures, Nuremberg doll kitchens and technical and mechanical toys were produced in the city and surrounding country and distributed by a very effective trade network extending far beyond Germany. The Frenchman Leo Claretie stated: "Nuremberg had the reputation of making the most beautiful and most modest toys that were produced in quantities such as to flood all of Europe."[6] This statement describes the wide distribution of Nuremberg toys--they could be found everywhere. In Bestelmeier's list of 1798 there were some 8000 "toys, artistic and other useful things" listed.[7] Just as remarkable as the quantity is the particular means of distribution through the Bestelmeier catalog; it was set up like the catalog of a present-day mail-order company, or like a collectors' catalog, with numbered, precisely described articles and many copperplate illustrations. It was more or less like a mail-order toy catalog of earlier times.
The great progress of the tinplate toy began at the time when machines were first used to shape the tin. These machines--presses and stamping machines--made it possible to press the body of the toy out of sheet metal much faster and in much greater quantities. A second prerequisite was the production of suitable sheet steel, for the sheet metal must not tear when being pressed. On the one hand, it had to be flexible and shapeable; on the other, stable and firm to withstand rough handling. A third and very important prerequisite was the elimination of the guilds' requirement of handcrafting.
Machine production began and mass-produced articles resulted. In Nuremberg the toy factories grew exceptionally well. At the German Industrial Exposition in Munich in 1854, eighteen Nuremberg and five Fürth firms displayed their wares. They showed particularly such metal toys as: children's trumpets, cookstoves, metal doll kitchens, kaleidoscopes, and weapons. Later on, rattles, musical tops, mechanical toys, railroads and magic lanterns were added, and finally, steam engines, microscopes and electric machines.[8]
Karl Rosenhaupt, who examined the Nuremberg-Fürth toy in-

dustry in a thorough study in 1907, shows by statistics that in 1875 there were 49 metal toy firms in Middle Franconia, employing 283 people, while twenty years later, in 1895, the number of firms had tripled--to 154--and the number of employees had quintupled--to 1671.[9] In 1907, in Middle Franconia alone, there were 476 metal toy companies, comprising 67.5% of the toy industry in the entire German Reich of that time. The Nuremberg-Fürth region was far in the lead in this business at the time.[10] These few statistics indicate the quick growth of this branch of industry which continued, although a bit more slowly, into the thirties; there was a definite interruption during World War I.

The trend toward the formation of large businesses with much machinery and large numbers of employees was obvious. But small businesses with two or three workers also existed in great numbers; they were founded quickly but usually disappeared just as fast. "The master of a small business was generally a trained metalworker. He took care of the soldering and assembly work. Stamping out the sheet-metal parts was done by his wife or female employees. The painting of the toys was almost always done as a cottage industry. The machinery that was used in manufacturing was operated by foot pedals. For pressing small parts, which called for the machines to be stopped often, foot power was in fact more practical. Pressing the metal parts was done by the master himself."[11] While the small businesses relied to a great extent on handwork and simple machinery, the medium-sized and particularly the large businesses quickly expended into machine production. Tinplate toys were made in great quantities on transmission-driven presses. The pressing of sheet-metal parts was done in a single work process. In it, a prepared sheet of metal was shaped on a single matrix. For example, the radiator grille, side panels, roof and rear of a car were made this way. These individual parts were then painted and assembled, soldered or otherwise attached, until the automobile was finished. L. C. Beck describes the working technology in a large factory, the tin toy firm of Matthias Hess, around the turn of the century: "All parts made of sheet metal, such as the sides, top and bottom pieces of the car body, the shaped parts of the powerplant, cannon barrels and mounts, etc., were made of pressed and painted or gilded sheets of metal, or stamped out of white tinplate in a form so only needed to be bent on bending machines and assembled or soldered on folding machines in order to produce the finished car body, locomotive body, cannon barrel or gun mount."[12]

As opposed to the tedious process of painting the tinplate which could be done only by hand, printing the sheet-metal parts (tinplate lithography), had long since been established by this time, as Beck goes on to describe: "For the car bodies, and to some extent also for the locomotive parts, sheets of tinplate are used on which the external features of these vehicles have been stamped in paint in the customary shapes and colors through a use of sheet-metal printing that replaces lacquering and painting. But many types of locomotives, that require soldering and thus have to be made of white tinplate, receive their paint job in the paint shop, which is equipped with drying ovens."[13] The soldering was done by gas soldering appliances, the wheels and hubs of the locomotives were produced by casting.

In the Seventies of the past century, the golden age of the Nuremberg sheet-metal toy industry commenced. Firms were founded whose names still remain very familiar to the collector, and whose products adorn many collectors' showcases today. The most important firms in the Nuremberg area, most of them founded before the turn of the century, are listed here in alphabetical order:

Karl ARNOLD since 1906
Hans BILLER (Hassi) 1935-1977-80
Gebr. BING 1866-1932
Josef BISCHOFF (Jobis) 1899-1938
BOEMER & SCHÜLER 1919-1974
Andreas (later Georg) BRANDSTÄTTER
(Geobra) since 1876, today PLAYMOBIL
Karl BUB 1851-1966
Georges CARETTE & Cie 1886-1917
Johann DISTLER 1900-1962
Peter DOLL & Co 1898-1936
Konrad DRESSLER 1918-1970
Gebr. EINFALT (Technofix) 1922-1977
Josef FALK 1896-1935
Georg FISCHER 1903-1958
Heinrich FISCHER & Co 1899-1931-32
Gebr. FLEISCHMANN in 1887
Martin FUCHS GmbH & Co 1919-1982
GAMA Georg Adam Mangold since 1882
Christian GÖTZ & Son (Göso) 1878-1960
Siegfried GÜNTHERMANN 1877-1965
Mathias HESS 1826 to circa 1941
Johann Emanuel ISSMAYER 1818-1935-36
KEIM & Co 1886-1960
Georg KELLERMANN & Co 1910-1979
Hubert KIENBERGER KG (HUKI) since 1907
Conrad KLEIN 1886-1913
Josef KRAUS & Co (Fandor) pre-1910-1937-38
Wilhelm KRAUSS 1895-1938
Ernst Paul LEHMANN since 1881 (in Nuremberg since 1951)
Georg LEVY (GeLy) 1920-21-1971
Johann Philipp MEIER 1879-1934-35
Josef NEUHIERL (JNF) since 1920, now CARRERA
Ernst PLANK 1866-1932
Gebr. SCHMID (Gescha) 1924-1967
Jean SCHOENNER 1875-1977
SCHREYER & CO 1912-1976
SCHUCO since 1921
Adolph SCHUMANN 1907-1939
Michael SEIDEL 1881-1982
L. STRENG & Co 1954-1965
TIPP & Co 1912-1971
TRIX founded in 1925 as Vereinigte
Spielwarenfabrike AndreasFörtner (ANFOE),
TRIX since 1938,
TRIX-MANGOLD (GAMA) since 1971
Fritz and Erwin VOIT 1919-1969
Heinrich WIMMER OHG 1928-1976
Fritz WÜNNERLEIN & Co (Wüco) 1906-1960
GEORG ZIMMERMANN 1946-1972.14)

This far-from-complete list of the tinplate toy firms in Nuremberg and the surrounding area--Fürth, Zirndorf, Stein, etc.--reads like a "who's who" of this industry. Nuremberg had become the center of the tinplate toy industry.
How this industry came into being over the course of centuries has already been shown, and why this industry attained such

significance in Nuremberg surely has a lot to do with the already noted Nurembergers' love for small objects. Thus it is no accident that the most famous Christ Child Market in the world takes place in Nuremberg, where thousands of little things can be admired and purchased from countless booths. The fact that Nuremberg is also the center of the German pencil and writing implement industry, with worldwide renown, also suits this characteristic. Even in the realm of metal goods have many small goods been produced: needles, screws, nails and springs. And last but not least, the food: the famous Nuremberg bratwurst sausages are the smallest of their kind in the world. Silly, but true. Small things are obviously dear to the Nurembergers, and carving and building, improving and developing are part of their nature. This probably explains the Mechanici, who passed on their skills for hundreds of years, tackled technical developments and became famous in Nuremberg as real innovators.
The toy industry needed these people, and still does: the master mechanic, the toolmaker, the model builder and--above all--the brilliant producer of ideas, inventor, designer. Heinrich Müller was such a man, and he ranks as one of the most imaginative tinplate toymakers. He is the man who made SCHUCO. Born in 1887, he tried to run his own business of making cars, torpedo boats and airships by hand at the age of seventeen. In 1909 he joined the BING firm, already world-famous, as a modelmaker. In 1912 he left there and, along with the merchant Heinrich Schreyer, founded the firm of Schreyer & Co.
When the name SCHUCO was introduced in 1921, no one suspected that a firm had been founded that would one day attain world fame. In the thirties, and particularly after World War II, SCHUCO cars were sold on the international market, and it was always Heinrich Müller who was the tireless innovator. He is regarded today as the most legendary of all Nuremberg toy creators.
One of the few Nuremberg men who can report authentically on the great era of tinplate toy production after World War II is Helmut Kellermann, heir to the Georg Kellermann Blechspielwarenfabrik located on Hessestrasse in Nuremberg. This firm produced tinplate toys until 1979, and was thus the last great Nuremberg firm in the field. Helmut Kellermann was not only the head of the firm, but also an originator, designer and technician. In an interesting account, he portrayed the development of a tinplate toy car from the idea to the series production. This report provides detailed insight into the technique and work of a tinplate toy producer and makes the laborious procedures of shaping sheet metal understandable. Thus it shall be reproduced here in comprehensive form.
When the decision was made to produce a new article, i.e. the Porsche 911, then Helmut Kellermann wrote to the manufacturer and asked that factory drawings of the car be made available. From these factory drawings, he then prepared a design drawing in 1/33 or 1/35 scale. These were the scales that were usually used for Kellermann cars. In the first great era of the tinplate toy in the days of the Bing brothers before World War I, automobiles, locomotives or other toys were never made as scale models of the original. The Model T Ford made by Bing, shown in this volume, looks very similar to Henry Ford's famous car, but is certainly not made to scale. Only since the Schuco products of the mid-thirties has the tinplate toy developed into a scale model. In the seventies, Kellermann produced only scale models, which required very precise design drawings from the start.
Helmut Kellermann needed about three weeks to prepare these drawings--which were made with the technique known as

Punktübertragung. The drawings then went to the toolmakers. To aid the toolmakers, patterns were made. Now the most important work for the later series production began, the production of the "drawing stamp," that tool with which the basic form of the automobile was made out of a pre-stamped piece of sheet metal on drawing presses in a single work process. The production of this drawing stamp took another three weeks. Kellermann worked without the building of a wooden model, working directly on the steel block instead. At first the shape was machined crudely out of the mass of metal. Then work continued, a tenth of a millimeter at a time, until the model form was finally finished. Photos of the original car were also used in the process. Every model had its "difficult points;" those of the Porsche 911 were the front headlight rim and the rear side panels, which included a concave curve. Here the work had to be done with emery paper, 1/100 of a millimeter at a time. When the drawing stamp was finished, the so-called drawing matrix, had to be made. From a second steel block, the toolmaker now carved out the hollow form with an inside cutting device. Then came the first attempts to press the model out of sheet metal. Often it happened that for the first few times only defective items came out of the machine. The mistakes were clearly seen at once too: in some places the sheet metal was too thin, or had torn, while in others it was too thick or had formed creases. In tests that took days, the work of making improvements had to be done. There was always the danger of cutting the block back too much, so that the form was "destroyed"--the weeks of work would have been in vain. Here it was up to the toolmaker's capability to attain the correct "fitting" for the tool after much testing. Many a toy car still had its problems, though. In the Porsche 911, the shape of the fender was problematic where it gave way to the headlights; the sheet metal tore there again and again. Therefore the original scale had to be departed from; the fender was inclined to the front a bit, so that it fell away more to the front. Finally the sheet metal "held" and the form could be used for mass production.

Pure sheet steel was used, such as is used in automobile production, but rolled thinner. It had a thickness of 0.24 mm and ranked in Quality Group VIII. Kellermann--and many other toymaking firms as well--ordered it from the Westfalenhütte of Dortmund, which was owned by the Hoesch steel firm. This was the first firm in the world that had produced cold-rolled sheet steel, circa 1930. This sheet steel had decided advantages over the earlier, warm-rolled steel: it was considerably more elastic and thus also more tensile, and it had a very smooth, glossy surface. Only with this steel did it become possible to make even difficult

The complete tools with distance plate no. 416 for the Ford 20 M Taxi, 1964.

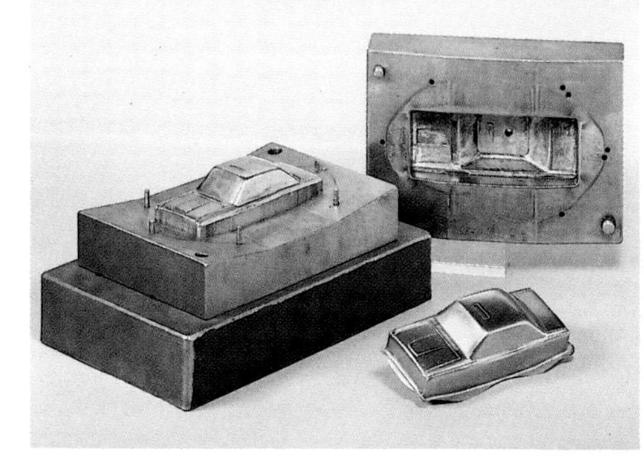

The original body for the Ford 20 M, G. Kellermann & Co.

model shapes in a press. The steel was delivered in panels measuring 533 x 765 mm. These "uneven" dimensions were derived from earlier British inch measurements. All the toymaking firms ordered sheet steel panels of this size.

The drawing tool produced the basic shape but, to make the model complete, usually ten to twelve other tools had to be produced: holes and sliding channels for windows, wheels and headlights, stamping tools for doors, bumpers, front and rear ends, cutters for other incisions into the metal. A tinplate toy car had to go through about twelve work processes before the complete basic shape of the body was finished. Then came the baseplate and finally the mounting of the motor, axles and wheels, seats, steering column and driver figure to the body, which had meanwhile been painted. Finally it was packed along with the steering controls--only then was the article ready to be shipped out. In the Sixties, Kellermann had 35 machines running, and up to 10,000 articles per day left the factory. Even when production was halted in 1979, there was still a large assortment in production. The decision was made that the salary costs could no longer be met by the company income. In addition, it had become extremely difficult to find qualified young workers for this type of work. The era of the tinplate toy had finally come to an end.

Many firms are no longer in existence. Today the firm of TRIX-MANGOLD in Nuremberg-Fürth still produces small quantities of SCHUCO replicas, since the demand for these toys is steadily growing among collectors. Of course, the old original models, especially in their original boxes, are most in demand. Still in all, Nuremberg is still alive as a toymaking city. The large firms that exist there--ARNOLD, BIG, FLEISCHMANN, LEHMANN, TRIX-MANGOLD, PLAYMOBIL, CARRERA, to name just the most prominent ones--have long since conformed to the trends of the times and produce modern plastic toys by die-casting methods, high-quality model trains in various scales and track widths, and fast racing cars. The names of SCHUCO and BING are now symbols of the golden age of the tinplate toy, living on in the hearts of collectors.

Footnotes:

1) Doppelmayr, 1730, p. 287, quoted in Wenzel, 1967, pp. 47-48.
2) Wackenroder, 1984, p. 122.
3) Wackenroder, 1984, p. 122.
4) Wackenroder, 1984, p. 52.
5) Bestelmeier, 1798, pp. 34ff, quoted in Wenzel, 1967, p. 49.
6) Claretie, n.d. (1894), p. 24, quoted in Wenzel, 1967, p. 89.
7) Wenzel, 1967, p. 92.
8) Wenzel, 1967, p. 49.
9) Rosenhaupt, 1907, p. 49.
10) Reher, 1977, p. 25.
11) Wenzel, 1967, p. 180.
12) Beck, 1899, p. 460.
13) Beck, 1899, p. 460.
14) List based on Christiansen & Leinweber, 1985, and Metzger, 1985.

Chapter 2.

THE SCHUCO FIRM

The story of Schuco is inseparably linked with the name Heinrich Müller. He was born in 1887; by the age of seventeen he had already developed his own ideas for toys and, along with his brother, was producing cars, torpedo boats, Zeppelins and airplanes. At the age of 22 he joined the firm of Bing, which was then world-famous, as a modelmaker. He stayed there only three years. In 1912, along with the merchant Heinrich Schreyer, he founded the toymaking firm of Schreyer & Co. In an area of about 150 square meters on Roon Street in Nuremberg, they began to produce "felt and plush toys."

Thanks to their good workmanship and original mechanisms, the articles immediately found a place in the offerings of dealers, and business was good. But World War I interrupted all hopes of expansion. The firm was reorganized in the spring of 1919 with another investor--the textile merchant Adolf Kahn. The new factory facilities were in the basement of the Wagner drive-belt factory at 26 Singer Street. Business gradually got back into action, and the German toy industry was able to regain markets that had been lost. In 1921 the new trade name of SCHUCO was introduced, a brand that was later to gain international fame. Heinrich Müller's mechanical toys were very much in the trend of the times; the moving, hopping, squeaking and trick-performing toy figure was the big seller in the shops. The most successful article in the twenties was "Pick-Pick," a sheet metal bird covered with soft plush; this toy was produced in unbelievable quantities--approximately twenty million!

Though the firm was already taking in money by the millions by the end of the twenties, and had moved to larger factory premises at 30-32 Fürther Street, the "SCHUCO Legend" actually began with the production of toy cars, which began in the mid-thirties. The "Wendeauto" (Turning Car) which never ran off the end of a table, became a sales hit, as did the "Studio-Auto," a simple name for the legendary Mercedes "SilverArrow" with which the German racing aces of the time--Lang, Stuck and von Brauchitsch-- drove from victory to victory. Not only the shape, but also the technology of this toy car was exciting. It had functioning geared steering, swing axles, free-running clutch, and crank starting, and even a differential which could be observed through a cutout in the baseplate. In the best of times, 8000 Silver Arrows--each consisting of 101 individual parts--were assembled every day.

The war brought toy production to a complete standstill, but in 1946 Nuremberg toys were already appearing inAmerican stores again. Heinrich Müller was 64 years old by now, but his energy and inventiveness seemed to be inexhaustible. Almost all the SCHUCO designs and patents originated from him personally. He also tested and improved "his" cars constantly. It has been said that he threw every model made by a pattern maker on the floor to test its durability. His products had to stand up to such treatment at the hands of children.

In 1952 a new factory building was erected on Fürther Street; the production premises now measured 13,000 square meters, and even that was not enough. With 800 employees, SCHUCO was Nuremberg's largest toy manufacturer. The hundred million toys sold in 1962 give impressive evidence of the worldwide distribution and sale of SCHUCO products.

Almost as fast as it had risen, the company declined not many years later. In 1966-67 there was a sharp drop in sales, and most of the tinplate toy production had to be given up. The attempt to convert to diecast metal and plastics came too late. Cheaply produced goods from outside Germany, and the new electronic toys, conquered the German market as they did others. The end of the tin toy era was also the end of SCHUCO, which declared bankruptcy once and for all in 1976. But today the name is a legend, and it lives on as such.

before 1930

since circa 1930

since 1980, used by GAMA (on Schuco replicas)

Schuco Wunderauto with ape, clockwork motor, three speeds, 1927. Nuremberg Toy Museum

The car goes forward, steered by the ape that turns the wheel. The ape's head also turns in the direction the car moves. The third "fast" gear makes the car skid.

Schuco Express Service Station with Auto, 1960. Fritz Ferschl Collection

Schuco Garage with Telephone 1500, Garage Auto 1750, prewar, circa 1935; pulling the telephone wire makes the car drive out of the garage. Fritz Ferschl Collection

Schuco Packard Synchromatic 5700, automatic push-button transmission, 1950. Fritz Ferschl Collection

Schuco Alfa Romeo with male driver, 1959, and with female driver, 1958.
Fritz Ferschl Collection

Schuco Grand Prix Racer 1070, 1954-1966. Ralf Regnat Collection

Schuco Studio-Auto 1050, 1936-1945; this best-selling Schuco car is based on the legendary Mercedes "Silver Arrow." Jürgen Franzke Collection

Schuco Radio Auto 4012, show-window model with visible interior parts (very rare example), as of 1952. Fritz Ferschl Collection

Schuco Radio Auto 4012, 1952-1964.

Schuco Ingenico remote-control car 5311, 1952-1960.

Schuco Limousine 1010, "the car that doesn't fall off the edge of the table", 1939-1964/65.

Schuco Elektro Police Car 5509, 1958-1969. Ralf Regnat Collection

Schuco Rollyvox 1080, 1957-1969.

Schuco Elektro-Phenomenal 5503, 1955-1969.

Schuco Examico I 4001, 1939-1959; center: Akustico 2002, 1939-1959.
Ralf Regnat Collection

The leaning figure can be dated as prewar (tools probably destroyed in the war); very rare models.

Schuco Akustico 2002, with removable "Schuco-Fritz" driver, ca. 1938.

Schuco Sonny 2005, the mouse swings the ball and waves while driving, 1952-1966.

Schuco Sonny-Peter 2006, moves arms and head, 1956-1960. Ralf Regnat Collection

Schuco Radio Auto 5000 with roof antenna, 1938-1943, one of Schuco's rarest and most technically advanced cars. Ralf Regnat Collection

Schuco Ingenico 5311, special edition, mid-fifties. Ralf Regnat Collection

Only a few of these nickel-plated cars were made and sold in metal-studded wooden boxes.

Schuco Go-Kart 1055, 1961-1968.

Schuco Elektro remote-control Ferrari 1073, 1973-1976.

Schuco Porsche Carrera RS, seventies. Ralf Regnat Collection

Schuco Curvo 1000, 1950-ca. 1954, makes circles and various other figures. Fritz Ferschl Collection

Schuco Elektro Construction Truck N 6070, 1949-1960. Ralf Regnat Collection

Schuco Elektro Construction Wrecker with Crane N 6070, 1956. Fritz Ferschl Collection

Schuco Elektro Construction Fire Truck 6080, 1965-1967. Ralf Regnat Collection

Schuco Covered Truck 6060 (so-called Hegi-Lkw), as of 1969. Ralf Regnat Collection

Schuco Truck, handmade chassis prototype (not put into production), 1950.

Schuco Dump Truck, handmade prototype (not put into production), 1950, with mechanical tailgate.

This swiveling crane was announced by Schuco in 1955 but never put on the market. The example shown here, dating from 1954, shows the last stage of development.

Schuco Swiveling Crane, 1954, which could be combined with the fire truck. Fritz Ferschl Collection

Schuco Lastomat 6084, ca. 1965-1970. Ralf Regnat Collection

Lastomat with electric motor for forward and reverse motion; with the switchable electric motor, the Meiller tailgate could be used to raise and lower small loads.

Schuco Bobby, 1954-1964, built for the British market, very rare in Germany. Fritz Ferschl Collection

Schuco Pick-Pick Bird, newly issued in 1928.

This very natural-looking "dancing, hopping and pecking" bird sold well for many years. Up to 42,000 pieces per day were produced. Delivery time took about two months in 1928.

Schuco Akrobato-Purzler, new in 1920. Nuremberg Toy Museum

Schuco Penguin 940, 1961.

He dances and turns on his own axis.

Schuco Walt Disney Little Pig 980/2, new in 1934; Clown 986/2, postwar; both playing violins.

Schuco Dog, ca. 1930.

"Who is leading whom here?" In this case, the shoe (with Japanese clockwork) leads the dog!

Schuco Dompfaff (bullfinch) Bird, standing, 1923 (rare). Fritz Ferschl Collection

Schuco Mickey Maus Twisti, 1963.

He and she lean left and right while moving forward.

Chapter 3.

A CHILD'S DREAM

From the Life of a Tinplate Veteran Matthias Murko

So this was to be his place for the next few weeks, behind glass, bedded on dark red velvet. From above came the glowing light of a bulb, the heat from which could not, of course, have any effect on his tinplate clothing, though it did make the plastic letters of his name fade very pale.

Day after day he was exposed to the sight of countless observers, and many children pressed their noses flat against the window glass. He could not rest until the lights were switched off and the alarm system turned on in the evening. Particularly in the long nights that were interrupted only now and then by strange sounds, by rattling and thumping, he sometimes had the feeling that he was not alone. It seemed to him as if, in the big room outside the glass pane, there were others of his kind enduring the same sort of existence. In these long nights, many memories came to him, going far back to his childhood almost forty years ago.

His birthplace was a tinplate toy factory in Fürth. There he received his body, baseplate and interior of drawn sheet metal, held together by metal strips that ran through slits and were bent over. His strikingly shaped bumpers and chrome trim always attracted attention and admiration, but the whitewall rubber tires, mounted on steel hubs with chromed metal hubcaps were his pride and joy. Unlike many of his expensive fellows, he had only a flywheel as a powerplant. A red and white paint job and precisely detailed interior made him one of the handsomest cars of his time. His resemblance to his big brother was amazing, a quality that scarcely one of his fellows could claim to possess. At first he envied those fellows who were equipped with an electric motor and remote controls. But over the years, this changed to a feeling of satisfaction, for he, with his simple powerplant, had outlived almost all of his friends and acquaintances.

After he had passed his manufacturer's stringent final examination with honors he landed, packed in a colorfully printed box, in the show-window display of a toy store in downtown Nuremberg. On a cold, windy evening, shortly before the store closed on December 20, 1954, he was very suddenly torn out of his opulent life in the show window and handed over to a middle-aged man, who took him in his colorful box, stuck him under his arm and left the store.

On the streetcar they traveled through Christmasy Nuremberg and on out to Mögeldorf. Then he could recall a short walk to a small row house with an even smaller front yard. By now the man had actually hidden him under his voluminous dark green coat. Stifling in the heavy cloth of the coat, he heard as if from far away the jubilant greetings of two children and the somewhat exhausted voice of a woman. He was hastily taken up a flight of stairs, and then he disappeared for a long time into the darkness of a closet. Hauled out of his pitch-black hiding place, his box, colorful as it was, was covered by even more colorful wrappings and tied, only to be hastily and completely unwrapped again some time later.

The ensuing hours were then the most important and most turbulent of his long tinplate-car life: accompanied by jubilant voices, he was liberated from his colorful wrappings by a little blond boy whose hands shook with excitement, and held in front of a small fir tree that stood in the room and sparkled with candles and strange ornaments. The boy clung to him the whole evening and then, disregarding the protests of the woman who was presumably his mother, and whose voice he had recognized at once, took him to bed. Clutched firmly to the boy, he went with him into the land of childhood dreams. This period of first love lasted only a few days; then he had to spend the nights in a big chest, along with several comrades and a crowd of the strangest objects. Big soft plush animals lived there, along with a great variety of blocks, cubes and stones, which gave his nice tin body its first dents and scratches. Only rarely was he able to sleep under the big, soft teddybear, who protected him from the attacks of the chest's other inhabitants.

With the beginning of spring, his range of activity expanded considerably. To be sure, his flywheel drive was not off-road capable, but this scarcely seemed to bother his owner. The boy took him along to a brownish-green surface called a field, though there was little grass to be seen there. On they went, through thick and thin, to a sandbox, crossing which was often extreme and not endurable to his inner organs. His exterior also came to look worse and worse. Without a doubt, that was the reason for his owner's ever-decreasing interest in him.

The appearance of a rival on the boy's fifth birthday resulted in complete indifference to him, all the more so as the rival was off-road capable and even had remote controls. But this development had one advantage, namely that he was ususlly left in peace and could recover from the strains of outdoor life. In the course of his infrequent playtimes, his inherently very robust flywheel drive and rear axle expelled all the bits of dirt and sand, and ran very well again. Up to this point his memory was very clear, as if it had been today. If there is any truth in the widely held opinion that memories of one's early years become especially vivid in one's old age, then he must have become quite old while resting in the toy chest.

Few events took place in his next years. His owner had long since grown out of his childhood when one day our tinplate car, along with a big bunch of other toys, appeared on a wobbly table at the Nuremberg Flea Market. In the morning haze of the second day it finally happened that he, along with five other tinplate fellows, all of whom he knew very well from his early years, were sold and taken away in a rattling delivery van.

This person from Fürth, it was said now, certainly did not want to keep him personally. And in fact, a few months later the man took him to a collectors' meet. Suddenly many people were interested in him; it almost seemed as if he were one of the most important representatives of his kind. Only after long and noisy dickering, though, did he acquire a new owner--at many times the flea-market price.

The new owner obviously took great pleasure in accompanying our tinplate friend with more and more of his kind. Perhaps that is just the reason why he had such a big house, in which he usually lived alone. The many rooms and chambers were stuffed full of the most varied objects. Enamel signs, boxes, and a great number of old clockwork-powered items completely filled the cellar. The living and dining rooms were filled with old toys of all kinds. Three disassembled Harleys had been in the kitchen for some months, and in the bedroom were thirty-five historical iceboxes. The other rooms seemed to be similarly occupied, but this was known to him only through hearsay.

For years, or even decades, he slumbered among the treasures of his somewhat strange possessor, until the man got him out one day without any warning. Now he was cleaned and polished; one of his plastic name plates was even replaced by an undamaged one that his owner had taken from a badly damaged comrade. So now he glowed again in his oldtime glory, and even a brightly printed box appeared again--everything was just as it used to be.

Then for a long time nothing happened, until one evening when his owner took him out of his corner of a cabinet, packed him into a big silver aluminum case, and took him on a long trip in the car. After something more than an hour, he heard, though somewhat muffled by the box he was in, hearty greetings and ensuing conversation, of which the world of tinplate things was the focal point. Suddenly the box was opened. He found himself on a sturdy oak table in a room at a hotel in the suburbs of Nuremberg, and seven or eight knowledgeable faces were bending over him and the six other tinplate cars that, as he now noticed for the first time, had also been in the box with him.

After a long and soft-spoken conversation, he was now taken out of the box, which was closed again immediately. Only when he was placed in a big padded case along with several other tin cars, all of which looked much more handsome to him, did he realize that he was being subjected to the most stringent criteria of several red-eyed connoisseurs. On the lid of the case, big letters spelled out "Selected Items for Display--Please do not tilt!" He was taken some time later, shaking and being tilted endwise several times, up several flights of stairs. Then white-gloved hands put him carefully into the already mentioned case. And now there he stood, thoroughly enjoying the obvious interest in him, the fuss over him--despite the heat of the glowing spotlights.

Weeks went by; the show had already been extended twice because of its enormous success, when one afternoon a middle-aged man stood before our tinplate veteran's glass case and stayed there for a good half hour.

It is hard to say who recognized whom first. The occupant of the case was surprised that this man stared at him for so long, and then he recognized his eyes again. They were the eyes of the small boy, of whose blond hair practically nothing was left. When a bittersweet smile brightened the face of the observer, it was clear that the two of them had recognized each other again--after more than forty years.

Chapter 4.
THE BING FIRM

The brothers Ignaz and Adolf Bing opened a distribution firm for household goods and toys in Nuremberg in 1866. The small merchandising firm flourished and the Bings decided, on the occasion of the Bavarian State Commercial and Industrial Exposition of 1869, to found their own manufacturing firm.

Just two years later the *Nürnberger Metall- und Lackierwaren-Fabrik, vormals Gebr. Bing* [Nuremberg Metal- and Lacquer-ware Manufacturing Co., formerly Bing Brothers] employed 100 workers, with just as many working at home. At the 1882 State Exposition the firm was already exhibiting the largest assortment of any manufacturer. They shipped to international markets overseas, as far away as the Orient. Six thousand articles were offered in their catalog, and they had by no means ceased to expand.

They produced tinplate and other metal toys, as well as metal household utensils of all kinds, from tea sieves to coffee services of sheet metal. In the *Illustrierte für die Blechindustrie* issues for 1912, the firm drew high praise. The catalog and price list was now 550 pages long; three lithography shops and four printers spent a year producing it. The firm had 2700 employees and had by now become the world's largest toymaking firm. Its offerings included just about anything that a child could want at that time: from steam engines to trains, from doll kitchens to the finest porcelain dishes, etc. Huge factory premises, occupied today by large sections of the Diehl firm, had been built on Stephan Street.

During World War I, it was necessary to convert to the production of military goods, and the number of employees kept growing; in 1915, 8000 people worked for Bing. After the war, the firm was changed to a stock company and finally, on account of its size, into several manufacturing facilities, each with its own director. With that--or so it appears today--began the firm's downfall, since the factories drifted away from each other. The decisive event, though, was the bankruptcy of the Bing sales firm of Concentra during the depression in 1929. The original firm had to meet the debts and also had to deal with the reduction in toy exports by two-thirds of its volume. That was too much, and in 1932 the bankruptcy proceedings began. After the forced reorganization of 1935, Ernst Schmidmer acquired 85% of the shares and formed the "Nowag" Noris-Werke AG, which still exists today under the name of "Noris-Tachometerwerke."

The Nuremberg firms of Karl Bub, Kraus and Fleischmann continued the toy production, but this succeeded for only a short time. Stephan Bing, who had formed a new company by the name of "Trix", had to flee from the National Socialists to Britain in 1938 because he was Jewish. The survival of the name "Bing" in Nuremberg at all is attributable to the entrepreneur Fritz Hintermayr who took over a branch of the company in 1932. The *Bing-Vergaserwerke* of today still brings back memories of this firm whose name was once world-famous.

Bing Mail Bus with driver figure, ca. 1925. Nuremberg Toy Museum

Bing Automobile with driver figure, one-off, ca. 1912-15. Nuremberg Toy Museum

Bing Model T Ford car, ca. 1925.

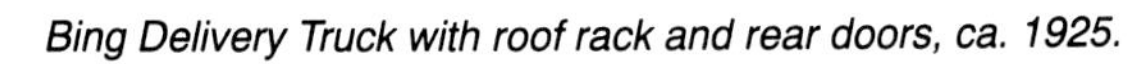

Bing Delivery Truck with roof rack and rear doors, ca. 1925.

Bing Limousine with driver, ca. 1925. Manfred Seeberger Collection

Left: Bing Open Car with driver, ca. 1912-1916; right: car with open driver's seat, ca. 1926-1930. Manfred Seeberger Collection

Bing Red Automobile, ca. 1930. Horst E. Schmidmer Collection

Bing Steam Fire Engine, ca. 1925.

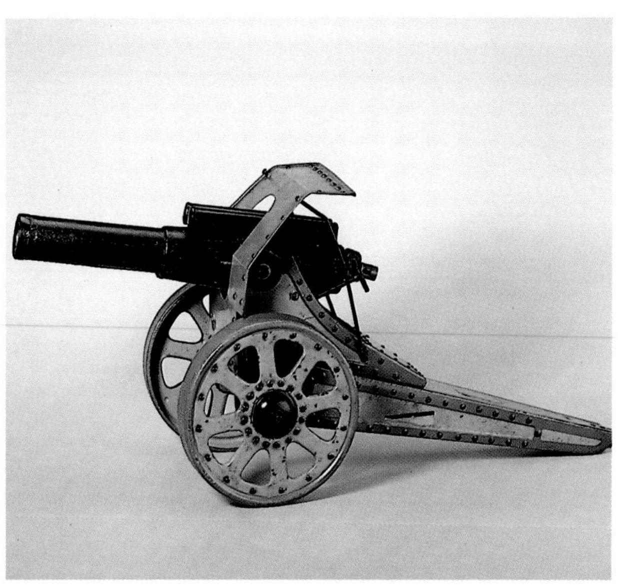

From this cannon, made shortly before World War I, peas can be "fired."

Bing so-called Pea Cannon, 1911. Manfred Seeberger Collection

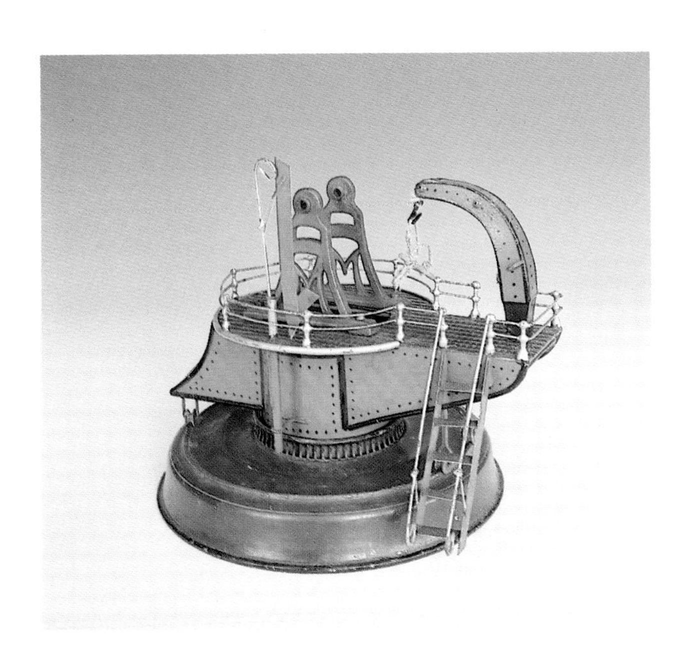

Bing Swiveling Crane, ca. 1912. Horst E. Schmidler Collection

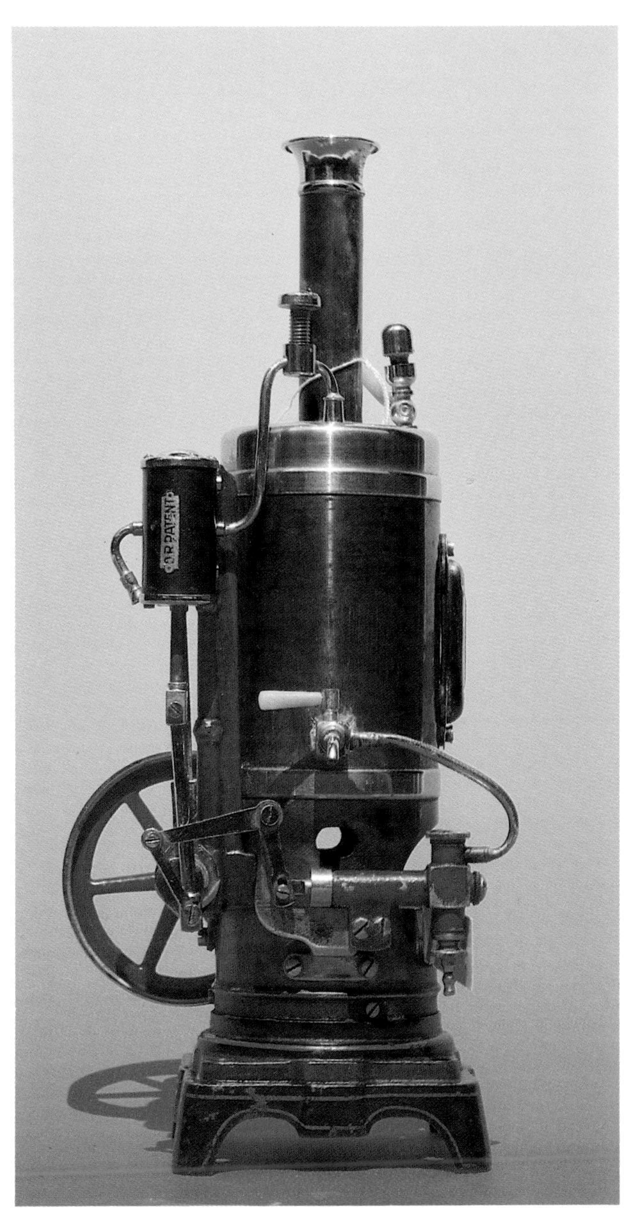

Bing Erect Steam Engine, ca. 1910 (built 1908-1925). Nuremberg Toy Museum

Bing Steam Engine with horizontal cylinder, large boiler and high smoke-stack, 38 cm high, ca. 1902.

Bing Steam Engine with vertical cylinder and boiler, generator and light bulb, 31 cm high, ca. 1906.

Bing Steam Engine with horizontal cylinder, vertical boiler, generator and light bulb, 25 cm high, ca. 1906. Manfred Seeberger Collection

Bing Drill Press, belt-driven, ca. 1905.

Bing Wood Saw with operator, ca. 1905.

Bing Blacksmith Shop (flat version), ca. 1912.

Bing Water Fountain, ca. 1930. Nuremberg Toy Museum

Bing Beer Brewery (one-off), ca. 1910. Nuremberg Toy Museum

Bing Wind-driven Mill, ca. 1922-1925. Manfred Seeberger Collection

Two Bing four-axle Express Locomotives, Gauge 1 (48 mm), ca. 1910 (built since 1902). Horst E. Schmidmer Collection

Bing cast iron Locomotive with clockwork motor, pre-1911.

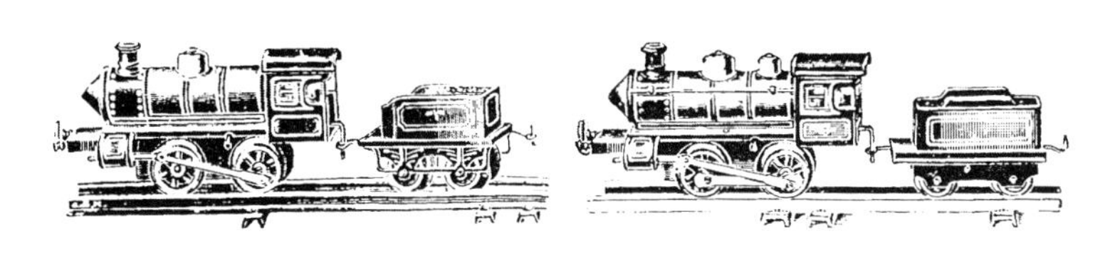

Bing two-axle Locomotive, clockwork motor, with tender and caboose, 1912.

Bing "King George V" Locomotive with clockwork motor, Gauge 2, British catalog, 1928. Manfred Seeberger Collection

Bing Roundhouse, 1909. Horst E. Schmidmer Collection

Bing "Elegant Depot" no. 11574, 1912. Horst E. Schmidmer Collection

Small Bing "Victoria Station" Depot, made for the British market, ca. 1920 (rare).

Bing Railroad Cars with observation balcony, "New York Central Lines", made for the American market, ca. 1920. Manfred Seeberger Collection

Bing four-axle "3-D Expresswagen" sleeping, dining, and mail cars, built for decades since 1902. Horst E. Schmidmer Collection

Two Bing freight cars for use with "Express-Lokomotive", Gauge 1, ca. 1912. Horst E. Schmidmer Collection

Bing Tucher Beer, milk, and Standard Oil cars, ca. 1930. Manfred Seeberger Collection

Bing Cogwheel Train with building, ca. 1928. Manfred Seeberger Collection

Bing Railroad Light, two-armed, ca. 1928. Horst E. Schmidmer Collection

Bing Harbor Crane, ca. 1928.

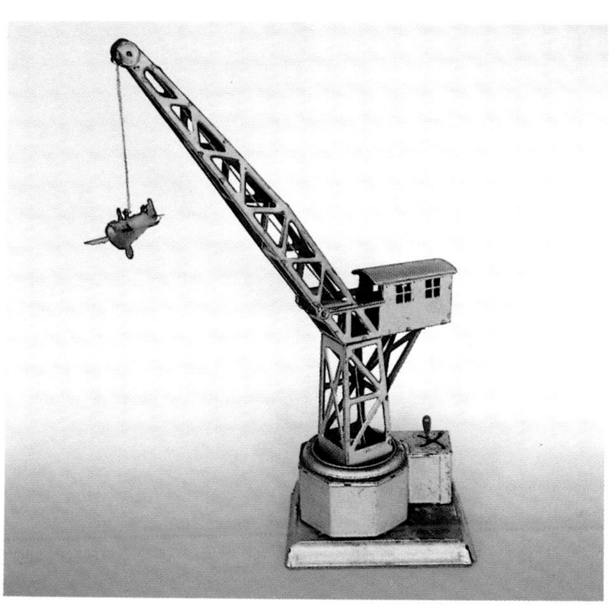

Bing Swiveling Crane with small airplane, ca. 1928. Horst E. Schmidmer Collection

Left to right: Bing Scale, 3 signs, handcart, ticket machine, signal, ca. 1910.

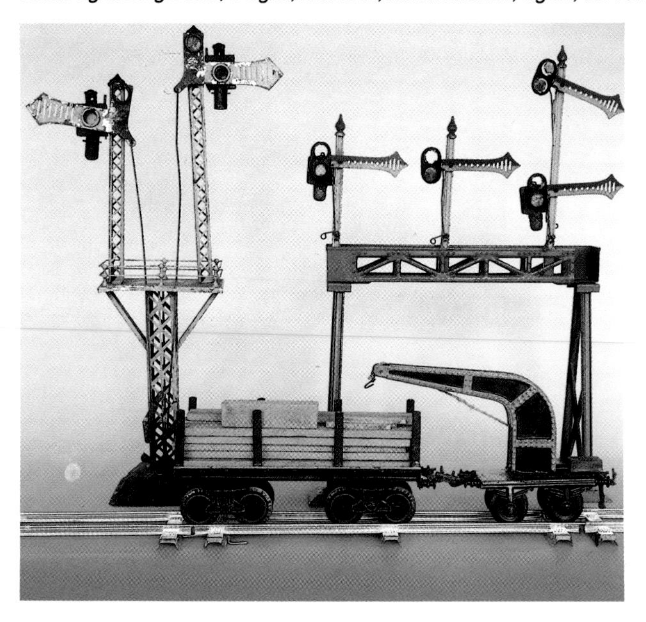

Bing Lumber and Crane cars, Gauge 1, with two signals, ca. 1910.

Bing Freight Depot with crane, 1902. Horst E. Schmidmer Collection

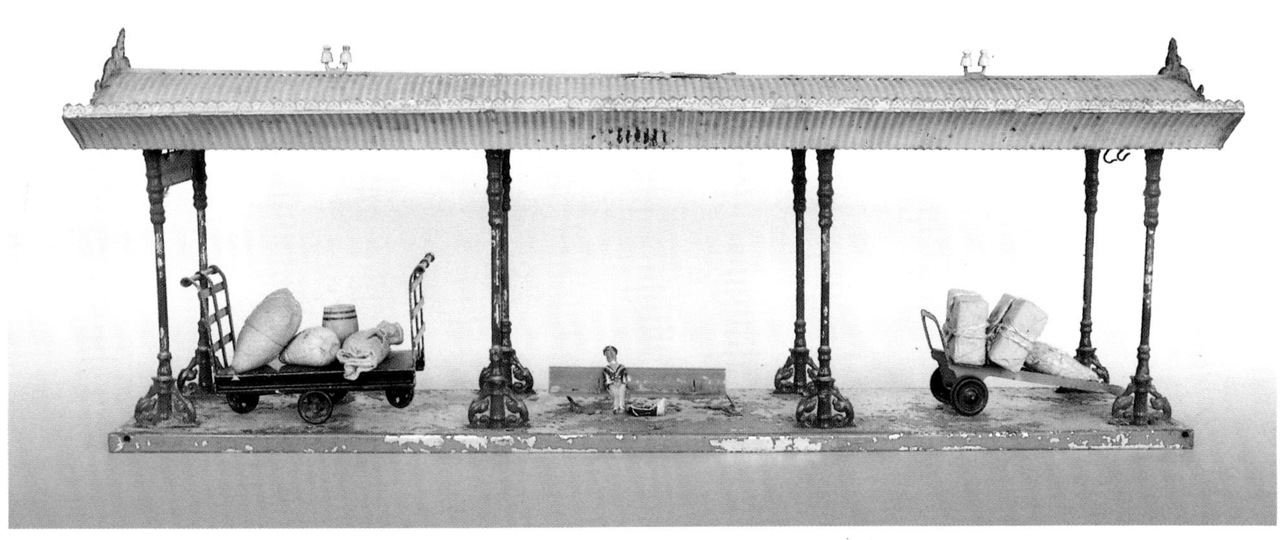

Bing Passenger BoardingArea with corrugated metal roof, ca. 1910. Horse E. Schmidmer Collection

Bing Crossing Guard's Hut with bell, ca. 1910.

Bing Crossing Guard's Hut with bell and gates, ca. 1910.

Bing Airplane, ca. 1928. Horst E. Schmidmer Collection

Bing High Speed Railcars in Zeppelin form, 1930. Manfred Seeberger Collection

Chapter 5.
OF DOLLMAKERS AND ANIMAL HORN TURNERS
On the History of the Nuremberg Toy Industry
Marion Faber

Like scarcely any other city in Germany, Nuremberg can look back on a long history of toymaking. The oldest witnesses to the production of toys in Nuremberg are small baked clay figures which were found during excavation work in the city in 1859. Numbering more than a hundred, most of them broken, they portray women, babies, saints or knights in armor. But there are also small pots, cans, bowls, hunting horns and other utensils, which were obviously used as toys by children. By the clothing of the figures it can be deduced that they were made in the late 14th century of light clay (because of its later use for tobacco pipes, it was also called pipe clay) with the help of forms or molds.

Figures and dolls as miniature reproductions of the human form rank among the earliest of all toys. In fact, the oldest documented trade that dealt with toys for children in Nuremberg is the *Dockenmacher*. The Middle High German word *tocke*, meaning doll, gave this trade its name. Some dolls were made of leather or cloth, filled with flax or wool, but the majority of dolls and toy animals were carved out of wood, painted or covered with fur and sold. These old *Dockenmacher* were thus dollmakers in the present-day sense, but also woodcarvers.

The word "*Dockenmacher*" came to mean toymakers in general in the 15th century, and soon other trades also dealt with the production of toys. Since the trades in the MiddleAges, and even until the beginning of the 19th century, were separated on principle according to the kind of material they used, no one trade could produce all kinds of toys. Despite the great differentiation among trades in Nuremberg in previous centuries, there were no specific toymakers; instead, most handcrafting trades made toys in addition to their primary products. Christoph Weigel reports in his book *Abbildung und Beschreibung der gemeinnützlichen Hauptstände* of 1698: "There is scarcely any handicraft in which that object that is customarily made in full size is not also often made as a small model or doll object for play." For example, in the 16th century at least seventeen handicrafts can be named which were involved in the production of toys: dollmakers and paper doll makers, wood and bone turners, carpenters, animal-horn turners, gold- and silver-smiths, beltmakers, jewelers, plumbers, alabaster smiths, bismuth painters, waxworkers, lead casters, candlemakers, tinsmiths, and bellmakers.

While at that time a carefully-made, expensive toy could be made only by the authorized trade, the production of small, cheap toys, of whatever material, was allowed for everyone as a "free craft." Since the boundary between the "organized" and "free" crafts of toymaking could not always be defined clearly, the result was that the established artisans made complaints to the authorities again and again in an effort to limit unwanted competition.

For the carpenter, one of the main fields of production was the making of expensive doll houses, which became a Nuremberg specialty in the 17th century and were contracted for by patricians and nobles. A trade almost unknown today, which developed in Nuremberg at the beginning of the 17th century, was that of the *Wildrufdreher* who turned out horns with which the sounds of various wild animals could be imitated. Since these articles, most of which were made of horn, became remarkably popular, the city government had to establish the *Wildrufdreher* as a "closed trade."

In the 17th century the turners, dollmakers, and alabaster workers in particular, turned more toward the production of less expensive and artistic but more popular toys, thus increasing their sales turnover considerably. Conversion to toymaking was so successful in Nuremberg during the 17th century that Christoph Weigel, in his book of 1698, could report that Nuremberg's toy production and exporting were so significant that the whole world was overflowing with its products. In the 17th century, Nuremberg had become a toymaking center in the truest sense, and was to develop even farther in this direction.

It was originally the production of wooden toys that won Nuremberg's fame as a toymaking center. Its favorable geographical position at the intersection of east-west and north-south trade routes made it an ideal commercial center. But it was not just the city's own toys that were exported all over the world, for the wooden and papier-maché toys from Berchtesgaden, Oberammergau, the Grödner Valley, the Erzgebirge and various places in Thuringia also passed through there. The products of these districts were bought up by wholesalers or middlemen and shipped out of Nuremberg to international markets. Nuremberg merchants maintained branches in the world's bigger cities and sold toys to even the most distant lands.

Although Nuremberg had, since the Middle Ages, actually been a city of the metalworking trades--the locksmith and tinsmith, gold- and silver-smiths, brass casters and jewelers--to name only a few--the production of toys was chiefly based on the material of wood until the first decades of the 19th century. Of course mechanical toys and artistic metal toys were made even before that, but metal was used in large quantities as a material for toys only for the figures made by the tin-casters. As a branch of the candlestick-making trade, there developed in Nuremberg in the latter half of the 18th century the occupation of the tin-figure manufacturer. The decline in the popularity of tin utensils resulting from the invention of porcelain was countered by many candlestick makers by the founding of *Offizinen* in which tin figures were cast.

The first well-known German *Offizin* was opened in Nuremberg in 1760 by Johann Gottfried Hilpert, who had come there from Coburg. After Hilpert, came several other masters who established traditional family businesses that were maintained for several generations. In the 19th century, Ernst Heinrichsen founded an *Offizin* which later became one of the most important in all

Three-piece doll house with kitchen, living room, and bedroom, 17th century. Nuremberg Toy Museum

Europe. Around 1848, Heinrichsen introduced the "Nuremberg size" or "Hinrichsen size" (28-30 mm) to unify the formerly varying sizes of tin figures, and this was soon accepted as the standard by other *Offizinen* as well. All imaginable subjects were portrayed in tin figures. In additional to military figures, there were also civilian figures, some based on natural history, others on the lives of city folk, farmers or foreign nations. Many graphic publications served as patterns. At the end of the 19th century, though, there was a definite emphasis on the production of tin figures of fighting knights, warriors and soldiers. Since the atmosphere of an epoch has a way of being reflected in the toys of the children, tin soldiers--before and during World War I--document a kind of national power.

Except for tin, metal toys were still the exception at the beginning of the 19th century. But then sheet-metal toys quickly came to the fore. The first sheet-metal toys made around the middle of the 19th century were, above all: magnetic swimmers, fountains, houses, chaises, trumpets, stoves, doll kitchens, kaleidoscopes, etc. In the last third of the 19th century, the list of wares became more and more extensive: rattles, noisemakers, trumpets, mechanical games, railroads, magic lanterns, steam engines, microscopes, electric machines, doll heads and many others. After the turn of the century, automobiles, airplanes and Zeppelins were added. Since about 1890, the industrial production of toys utilized colorfully printed tinplate. Instead of painting by hand or pattern, flat sheets of tinplate were printed in color by a lithographic process in series, and only then stamped out or pressed in a mold. On such a plate, many identical parts could be printed, thus achieving a high economic utilization of the material. Since not all toy manufacturers could afford to have their own printing presses on the premises, many contracts were given to specialized sheet-metal printers. Lithographed sheet-metal toys did not simply replace the hand-painted ones right away. Until the 1920s there were still individual manufacturers who painted their sheet-metal products in the traditional manner, but here too, the trend toward simplification set in. They converted to spray-painting, which guaranteed a particularly smooth and glistening surface. The individual parts of a sheet-metal toys were no longer soldered now, but rather mortised.

Whatever inspires the people of an era soon makes an appearance in its toys as well. Since mechanical and technical toys could be sure to gain the interest of an ever-growing number of customers, more and more firms were established in Nuremberg, Fürth and Zirndorf, and some of them were able to expand into large-scale industries. In this respect, the Nuremberg firm of Bing can be cited in particular as achieving not only gigantic production figures of its own, but also selling the products of other manufacturers in great numbers.

In the foreword of his book *Das Spielzeug im Leben des Kindes*, which sketches a panorama of children's toys at the turn of the century, Paul Hildebrand wrote: "For Germany, quite aside from the recreational and pedagogical characteristics of the toy, the preservation and study of them is a question of honor. While France and Britain, because of their more organized and undisturbed cultural development, have progressed beyond us in almost all branches of commercial art, and the modern artistic trends that move the world habitually come to us from precisely those countries, still in all we are far superior to those two nations in industrial terms, for since centuries we have been the unexcelled producers of toys, and our toy exports, which were already flourishing in Nuremberg in Dürer's lifetime, has now attained the outstanding level of circa sixty million Marks per year." In addition to the large and medium-sized firms, there were also many small family businesses, in which sheet and other metal were worked into all possible shapes. These small businesses still produced much by handwork and could convert quickly to new products when their customers' tastes required it. Shortly before World War I, the high point of economic development had been attained. The loss of the war and the political and economic conditions after 1918 had an extremely unfavorable effect on the Nuremberg-Fürth toy industry, oriented as strongly as it was to exporting. The German toy manufacturers encountered keen competition from American, British and French toys outside Germany. After the depression, the Japanese also pushed into the international market with their tinplate toys. Yet the legendary fame of Nuremberg toys was maintained, for they were still regarded as "quality products" in comparison to foreign products, even though this could no longer be true of mass-produced articles. What was included in the concept of Nuremberg toys in the twenties, other than the dominant sheet-metal toys, is cited by Ernst Wening in the December 1930 issue of the *Deutsche Spielwaren-Zeitung*: "Nuremberg-Fürth is also the capital of Germany's tin-figure in-

Doll kitchen of sheet metal, circa 1830. Nuremberg Toy Museum

dustry. In addition, an outstanding cloth-toy industry has developed there. The celluloid toy industry there is also significant. But there are also firms making wooden toys, artistic dolls, children's vehicles and bicycles, paint boxes. Another quite extensive industry is the manufacture of games, picture and paint books, as well as cardboard toys. The Nuremberg assortment may well include all kinds of toys with the exception of porcelain dolls, as well as all price ranges. It is, in fact, truly universal it terms of both types of toys and price ranges."

With the outbreak of World War II, the whole German toy industry came to grief, and particularly the part of it that used metal, because the materials were needed for war production. Many a toymaker could not or would not oppose the military trend that dominated education and toys. Whether this was out of patriotism or the economic necessity of taking part in the trade in tanks, guns and other military toys, still remains to be determined.

The period after World War II brought years of great change for the Nuremberg toymakers. The firms' postwar production was concentrated at first on a return to the successful tinplate vehicles, figures and other items from the thirties. Yet the end of the "tinplate toy era" could be foreseen. What had dominated the metal toy industry in Nuremberg up to that time had to deal now, in the fifties, with new plastics, and production facilities had to be revised bit by bit. The trend turned, and still remains, toward accurate scale models, particularly of automobiles and trains. For this type of production, new processes, such as pressure diecasting of metals and plastics, were far better. The production of the first diecast metal toys used almost exclusively tin and lead alloys. Because of the poisonous nature of these materials, the manufacturers later turned almost completely to zinc. The breakthrough for plastics casting took place only with the world's first injection plastics casting machine, which was developed and built in Nuremberg in 1956. Its design principles have not been equaled to this day.

Today most of the toy-producing firms in Germany are centered in Middle and Upper Franconia, around Nuremberg. The International Toy Fair held annually in Nuremberg is a showcase of practically the whole world's toy production.

MAC 700 motorcycle with clockwork motor, ca. 1949-1959. The driver moves up and down independently (sought-after collectors' item). Ralf Regnat Collection

Arnold Motorcycle 643, with sparkling light, clockwork motor, ca. 1935-1954. Ralf Regnat Collection

Crane Truck (rare), truck by Arnold, crane by Fuchs, ca. 1970; Bulldog Tractor no. 7300, fifties. Wolfgang Dittrich Collection

Arnold Caterpillar Tractor no. 7000, fifties.

Set of no. 3032-34-35-36, MAN diesel truck with four semi-trailers, 1956.
Wolfgang Dittrich Collection

Carette working model of an iron hammer, ca. 1906.

Carette working model of a water wheel, ca. 1906. Manfred Seeberger Collection

Carette two-axle Locomotive with tender, ca. 1904.

Carette Steam Engine, 1895-1900. Manfred Seeberger Collection

Distler Fire Ladder Truck, no. 3701, ca. 1928. Manfred Seeberger Collection

Distler Elektromatic 8000 side tipper, ca. 1960. Wolfgang Dittrich Collection

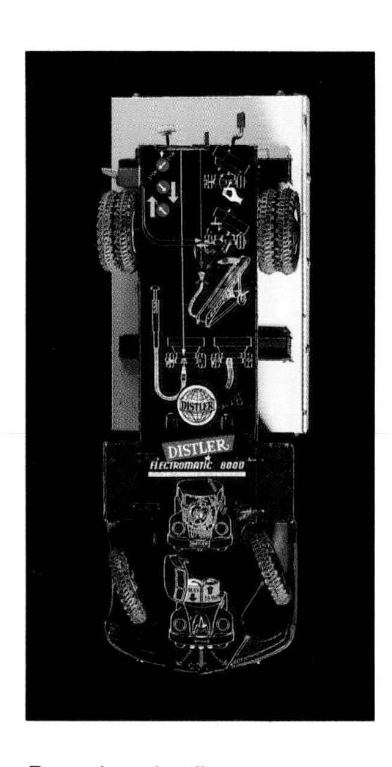

Baseplate detail

Distler Porsche Electromatic 7500, 1955-1962.

The legendary "Distler-Porsche" again, with original box and underside. Wolfgang Dittrich Collection

Distler Wanderer with stick shift, 1948. Klaus Seufert Collection

Doll & Co. working model of a water pump with two figures, thirties.

Doll & Co. working model of a musician, thirties.

Doll & Co. working model of a drummer, thirties.

Doll & Co. working model of a scythe-smith, thirties. Manfred Seeberger Collection

Doll & Co. Windmill, thirties. Manfred Seeberger Collection

Doll & Co. vertical steam engine, circa 1910 (firm in business 1898-1939). Nuremberg Toy Museum

Doll & Co. Elevator, thirties. Manfred Seeberger Collection

Doll & Co. working model of a bobsled run, ca. 1926. Manfred Seeberger Collection

Dressler Caterpillar Tractor, sixties (very rare). Wolfgang Dittrich Collection

Falk Steam Engine with vertical cylinder, 1925. Nuremberg Toy Museum

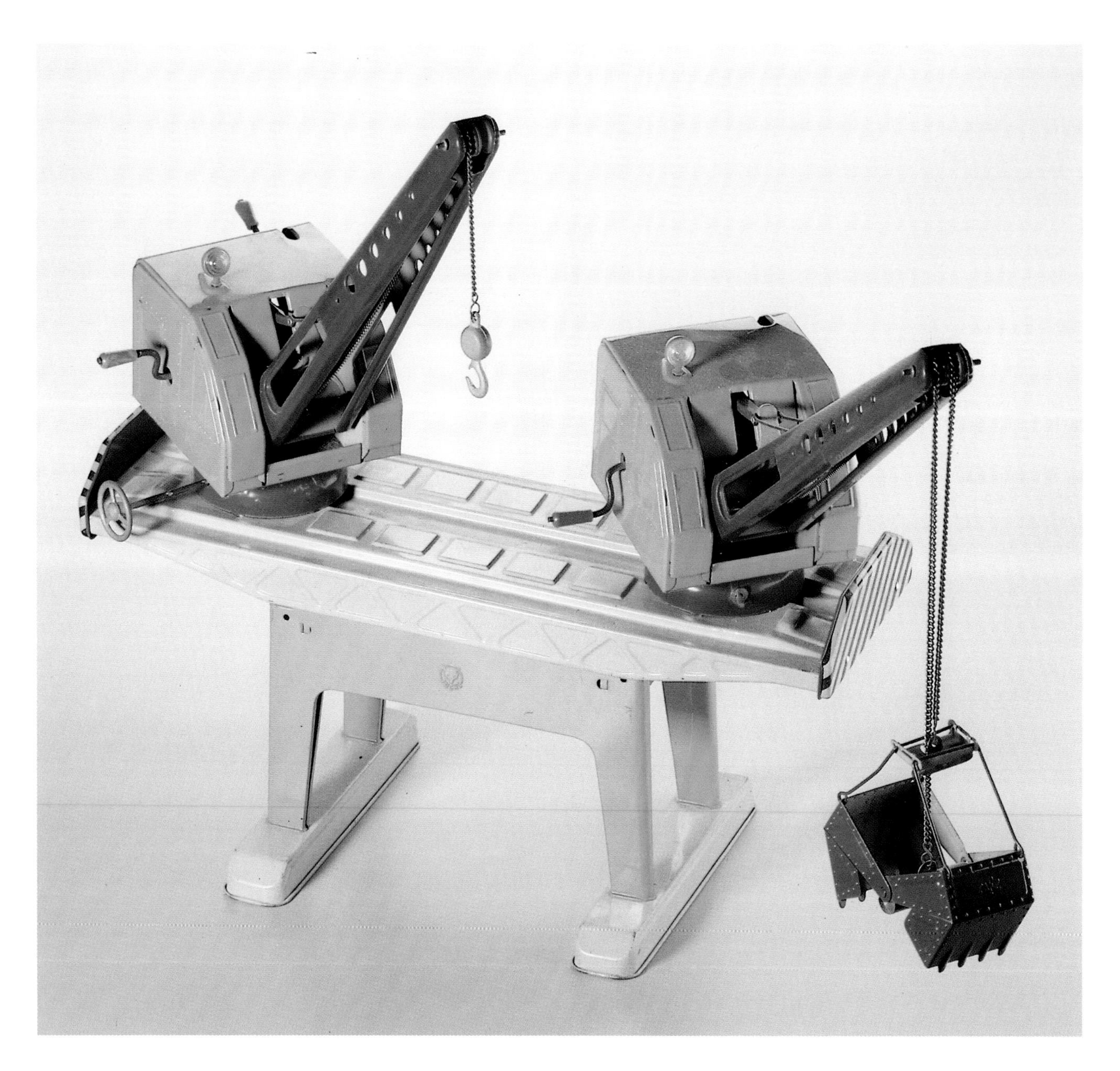

Fuchs Double Crane (display model), circa 1960. Wolfgang Dittrich Collection

Fischer Large Sedan, twenties. Nuremberg Toy Museum

Fischer Truck with open cab, ca. 1910. Manfred Seeberger Collection

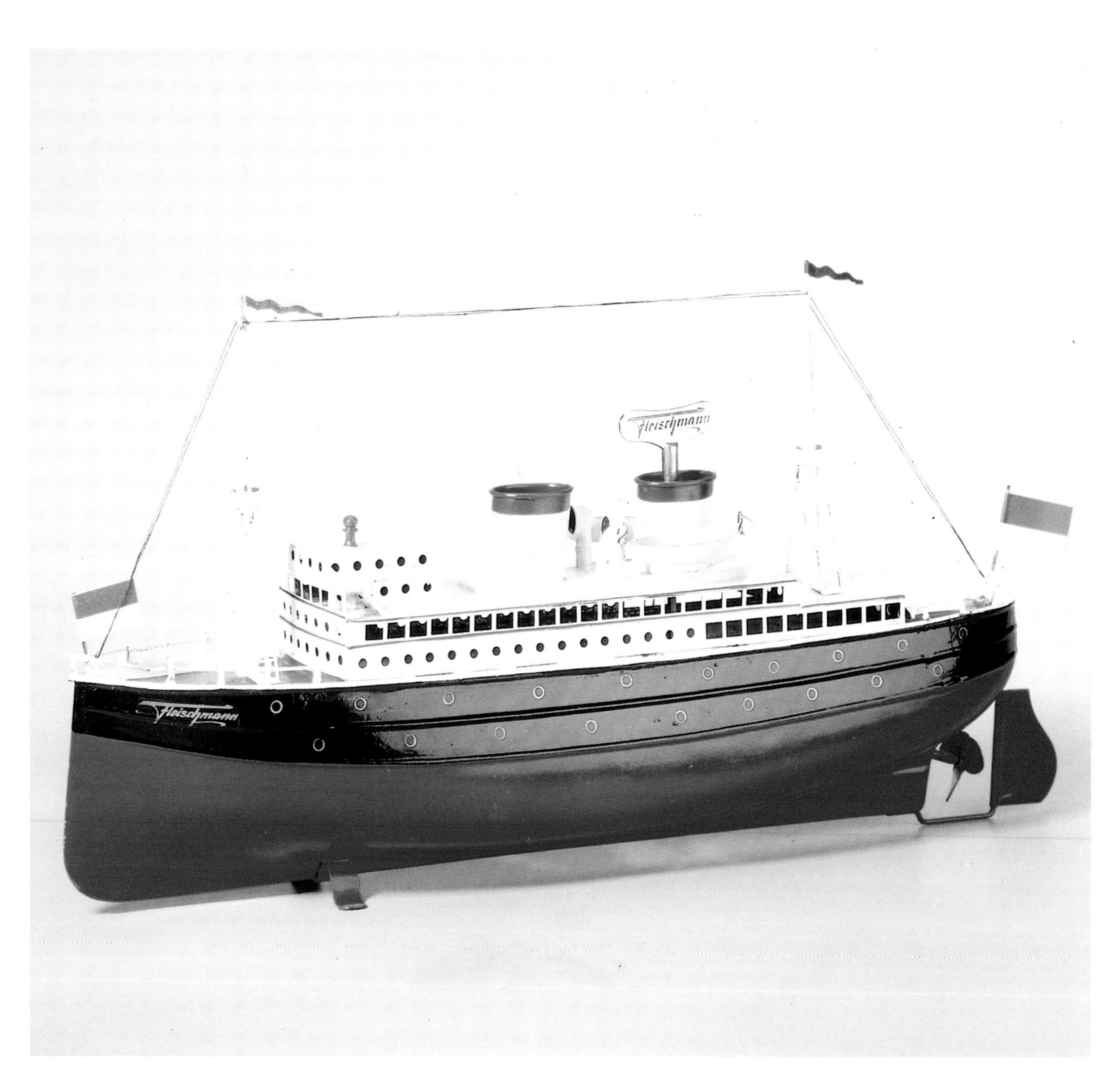

Fleischmann Passenger Ship with clockwork motor, 42 cm long, fifties.
Manfred Seeberger Collection

Fleischmann Road Roller, steam-driven, so-called "steam roller", ca. 1950.
Nuremberg Toy Museum

Fleischmann Opel Kapitän 400 with flywheel drive, ca. 1955-1965.

Fleischmann Cadillac with flywheel motor, ca. 1955-1965.

Fleischmann Opel Kapitän Police with flywheel motor, ca. 1955-1965.

Fleischmann Mercedes Police Car no. 104, electric motor (battery in trunk), 1958-1968. Ralf Regnalt Collection

This car turns at every obstacle. When going backward, the driver's head turns to the rear (very rare item).

Gama Police Motorcycle no. 126 with driver made of "Masse" (sawdust and glue), flywheel motor, and siren, ca. 1950. Ralf Regnat Collection

Gama Mail Truck, ca. 1966-67, and Bulldog Tractor (made in very many variations), ca. 1936-1972. Wolfgang Dittrich Collection

Gama Small Bulldog Tractor with Trailer, no. 776, ca. 1960. Klaus Seufert Collection

Gescha Combi 3 construction set, truck with various bodies, ca. 1955.
Wolfgang Dittrich Collection

Gescha Roaring Supercharged Mercedes Racer, 1938. Klaus Seufert Collection

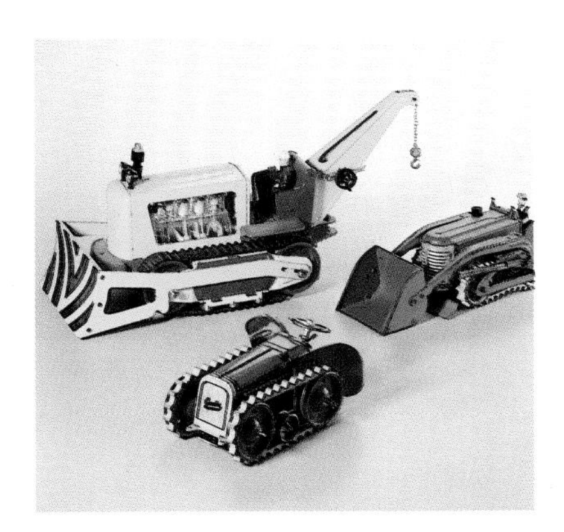

Gescha Caterpillar Tractors. Front: no. 71-2, thirties. Right: no. 720, early fifties. Back: no. 750, sixties, with plastic 4-cylinder motor. Wolfgang Dittrich Collection

Close-up of no. 71-2. Wolfgang Dittrich Collection

Geobra Telephone, 1937; Globe, 1939; Telegraph Set, 1939. Georg Brandstätter Firm Archives

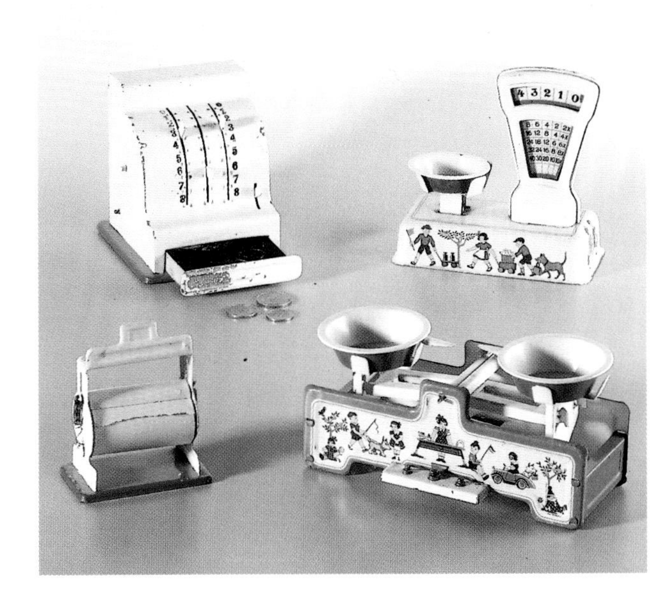

Geobra Cash Register, two scales with colorful designs, and paper roll, 1937.

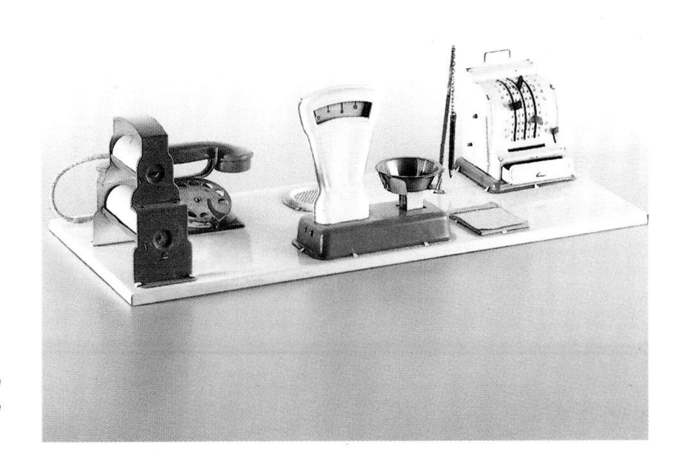

Geobra Store Counter with paper roll, two-color telephone, scale, cash register, change plate, note pad and pencil holder, 1937. Georg Brandstätter Firm Archives

Günthermann Road Roller, 1968-70. Wolfgang Dittrich Collection

Günthermann Fire Truck with man on ladder holding nozzle, fire chief and hydrant, thirties. Manfred Seeberger Collection

Günthermann Tour Bus with opening roof and original box, fifties.

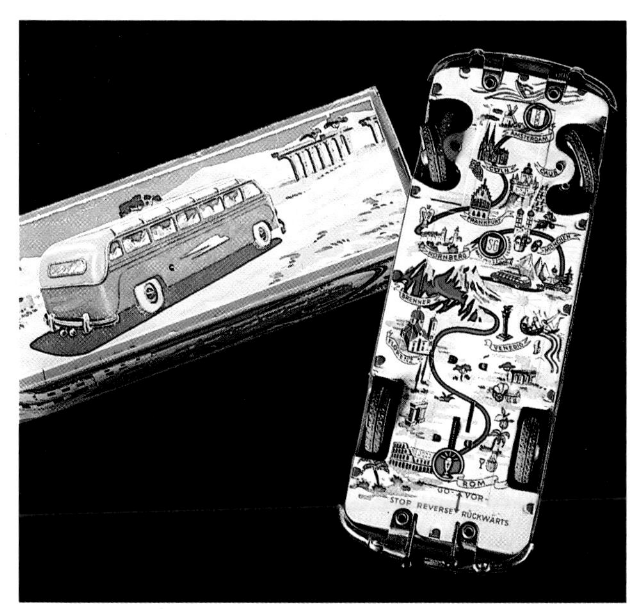

Detailed baseplate with tour route. Wolfgang Dittrich Collection

Göso Milk Truck and Fire Truck with floodlight, fifties. Wolfgang Dittrich Collection

Left side: Hess Fleet, consisting of one main ship and five to follow, ca. 1920. All ships are so-called ground-runners, thus cannot float. Right side: Two Hess Warships with guns and one Submarine, ca. 1920. Manfred Seeberger Collection

Left to right: JNF/Distler Telecar, 1954-60; Distler-Packard, ca. 1954; Mercedes 300 Sedan, ca. 1954. Wolfgang Dittrich Collection

JNF Silver Arrow "single-seat" racing car with electric power, ca. 1954 (rare oversize version). Ralf Regnat Collection

JNF Mercedes Racing Car no. 79, ca. 1954. Ralf Regnat Collection

JNF Battromot Mercedes 300 with electric power and headlights, ca. 1954. The car is steered via the star on the grille. Ralf Regnat Collection

JNF "Perfect" Automobile, 1954. Klaus Seufert Collection

JNF Porsche prototype with clockwork motor and lights, ca. 1954. Ralf Regnat Collection

When the fog lights are pushed in, the front hood opens. The headlight batteries are inside.

JNF Gigant "KE" with electric headlights and trunk, ca. 1954. Ralf Regnat Collection

JNF Indicator, ca. 1954, equipped with gearshift, steering, gas pedal, signals and ignition lock! Ralf Regnat Collection

Kellermann Cabrio Super, changeable car with mechanism to raise and lower the top, ca. 1956-1960. Ralf Regnat Collection

Kellermann Mercedes Fire Ladder Truck no. 439, 1972.

This truck no longer had an old-type baseplate but rather, for the first time, a new type of frame construction of stamped metal pieces.

Kellermann Mercedes Quarry Dumper no. 428, 1966.

Kellermann Mercedes Covered Truck and Trailer no. 431 A, 1968. Georg Kellermann & Co.

Kellermann VW Transporter no. 398, first issued in 1959.

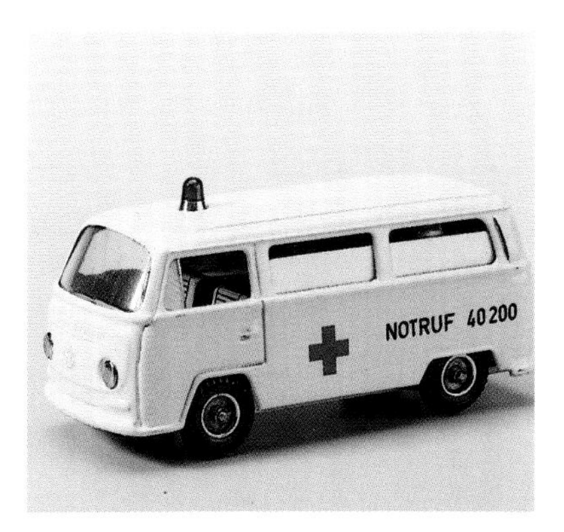

Kellermann VW Ambulance no. 402, first issued in 1960.

Kellermann VW German Mail Car no. 425, 1965. This model cost only 1.95 DM then! Georg Kellermann & Co.

Kellermann Mercedes 350 SL Convertible no. 447, 1978. The last model made by the Kellermann firm. Georg Kellermann & Co.

Kellermann Mercedes with Trailer no. 440/43, 1975. Georg Kellermann & Co.

Kellermann Montana Caterpillar no. 365, 1950. Wolfgang Dittrich Collection

Kellermann Tractor with Trailer no. 389, 1957. Georg Kellermann & Co

It is particularly noteworthy that the circus horse without a rider, which turns on its own axis, and the white cat, made of bakelite and thus fragile, have been coveted collectors' items for a long time.

Assortment of Köhler windup models made at various times. Manfred Seeberger Collection

The pretty songbird not only opens its beak but also moves its head, wings and beak while it twitters musically.

Köhler Singing Bird with clockwork motor and sound bellows, ca. fifties-sixties. Ralf Regnat Collection

Lehmann Dancing man, 1914-1945. Manfred Seeberger Collection

Lehmann Stubborn Donkey, 1897-1938; Uncle, 1903-1935. Manfred Seeberger Collection

When the car moves, the servant turns the parasol and the "Uncle" takes off his top hat.

Neidermeier Fire Truck, ca. 1962-64. Wolfgang Dittrich Collection

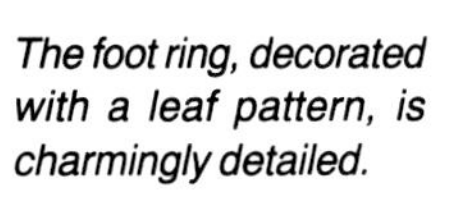

The foot ring, decorated with a leaf pattern, is charmingly detailed.

Plank "Ideal" Steam Engine with upright cylinder, ca. 1912.

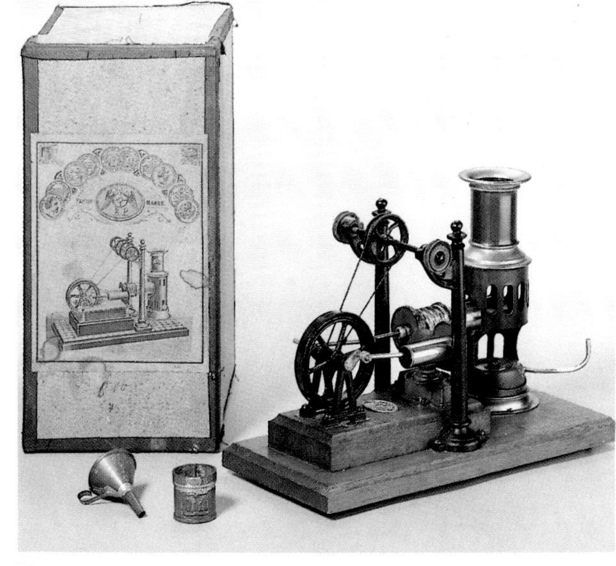

Plank Hot-air Machine with original box, ca. 1912. Manfred Seeberger Collection

Plank Traction Engine, ca. 1912. Manfred Seeberger Collection

Plank Steam Engine with horizontal cylinder, ca. 1905. Nuremberg Toy Museum

Seidel Police Speedboat, ca. 1950-60.

Seidel Police Boat, ca. 1950-60.

Seidel Steamship MS Rex, ca. 1950-60. Klaus Seufert Collection

Seidel Mercedes 190 SL Police Car, 1955-1960.

Seidel Mercedes Truck with automatic dumper, ca. 1960. Klaus Seufert Collection

Strenko Rocket Launching Pad, 1950-60.

Strenko Excavator, 1950-60. Klaus Seufert Collection

Technofix Helicopter with friction motor, 1958. Klaus Seufert Collection

Technofix Automatic Shovel Loader, 1958. Klaus Seufert Collection

Technofix Tractor no. 287, ca. 1960. Wolfgang Dittrich Collection

Technofix Tractor with Plow, 1957. Klaus Seufert Collection

Tipp & Co. Fire Ladder Truck no. 9590 with water pump and hose, ca. 1936. Manfred Seeberger Collection

Tipp & Co. Limousine no. 9502, ca. 1936. Manfred Seeberger Collection

Tipp & Co. Limousine with electric headlights, ca. 1936. Manfred Seeberger Collection

Tipp & Co. Military Police Car, ca. 1955. Klaus Seufert Collection

Tipp & Co. Heavy Road Roller, 1966-1968, goes forward and backward, the driver turning in the direction of movement. Wolfgang Dittrich Collection

Tipp & Co. Motorcycle with Sidecar no. 698, 1936-1945 (also made as a military model and in eight other versions).

Tipp & Co. Large Motorcycle with two riders, no. 597, very valuable, with clockwork motor, 1928-1945.

Tipp & Co. Motorcycle with Sidecar no. 59, ca. 1955, clockwork motor with on-off switch in sidecar.

Tipp & Co. Police Motorcycle no. 598, 1954-1964, flywheel motor, with mechanical siren and electric headlight. Ralf Regnat Collection.

Tipp & Co. VW Coca-Cola Transporter no. 020, 1954, with plastic cases in the cargo space, flywheel drive. Wolfgang Dittrich Collection

Wüco Road Roller (US Zone), 1954-1956, a rare model. Wolfgang Dittrich Collection

Chapter 6.

DESIGNS ON TIN

Ursula Kubach-Reutter

As always, the small boy's birthday at the beginning of November in the fifties was cloudy and rainy. But he knew that his father would once again give him something nice. This time it was a green car with red upholstered seats, all made of lacquered sheet metal, with rubber tires. First he looked it over very closely. Of course he could not read everything yet, but he tried to make out every letter, every number on the car, for they were very mysterious to him: "MADE IN U.S. ZONE GERMANY," "Schuco-Akustico 2002," and even "SCHUCO-PATENT-AUTO 2002." On other items made of tinplate there was lettering like "D.P.a." or "D.R.G.M. 1648602." Such toys were always something special, for his father had explained that these symbols referred to inventions and protection from illicit copying. That sounded very adventurous. Even today, as an adult, every time he passes through the flea market he examines every tinplate toy that attracts his attention for these ominous symbols. The various lettering appears on different parts of the objects and obviously has nothing to do with the functioning of the object as a toy, or anything of the kind. Some of them are hidden away on the underside, but on vehicles they are often placed very openly on the front or back, like a license plate, or even on the driver's door of a truck as if, for example, the firm of GAMA were a trucking company. Some lettering was stamped into the metal, some stamped or lithographed onto it, and some was on stick-on paper labels or riveted-on metal plates. As many as three or four different lettering systems could be applied to one vehicle or one steam engine: markings of the company, consisting of its business name or trade mark, a firm's logo, or an appropriate pictorial design, various protective markings such as "D.R.P.," "D.R.G.M.," "D.R.P.A.," "D.E.P.," "G.M.," often followed by long numbers. There were also indications of origin such as "Made in Germany," "Made in US-Zone," "Foreign Product," "Made in Western Germany," etc. In addition, there were often whole rows of mysterious numbers to be found, which could be either company, production, or even article numbers.

For many years, collectors and scholars have been trying to solve one of their biggest problems in studying tinplate toys with the help of these markings, namely the identification of the manufacturer and date of manufacture of an object.

Other aids to dating, which are of use in other areas, have proved to be unsuitable in regard to toys in general and tinplate toys in particular. For example, comparisons of style and architecture can be used only under strict limitations. Very definite influences, such as the necessity of suiting the item to use by children, particularly have to be considered. Nostalgic replicas (such as the Oldtimer Series made by Schreyer & Co. of Nuremberg) have to be kept in mind. Another imponderable, in the use of stylistic criteria in Nuremberg, is the fact that, under certain conditions, designers and modelmakers might change employers several times or found their own firms, without changing their personal style.

The question of technology and material is also very seldom of much help in and of itself, for as a rule, these criteria only indicate a time period before which something could not have been manufactured, but they very seldom pinpoint the actual time of manufacture. In model toys too, the production time of the actual object gives only a hint of this kind. The production date of the original and the model could, of course, be very different from each other. The firm of Schreyer & Co. in Nuremberg produced the Mercer Type 35 J only after 1960. Thus owners of this Mercer Type 35 J model do not own a model dating to the days before World War I, even though the original Mercer Type 35 J was made in that era.

At the present time, these designs on tin constitute the most definitive means of identifying tinplaye toys. Therefore it is necessary to acquaint oneself with their nature and their actual meaning by asking, "*Why* did *who* put this mark on *where*?"
Early tinplate toys very seldom bear identifying marks, although the principle of marking porcelain, glass or tin must have been known by then. But there were no legal or guild requirements that made identifying marks mandatory, as with tinware, nor was it in the best interests of the manufacturer, distributor or dealer to document the origins of tinplate toys as it was with glassware and later with porcelain. Nor did their origins yet indicate any particular level of quality.
All of this changed decisively, though, in the 19th century. The further development of the distribution system, the industrialization of commerce, and the technological developments in manufacturing led to complex changes in tinplate toy production. New technical developments required the deliberate use of more and more exciting, attractive ideas in toys, so as to put them on the market economically. The constantly growing application of ideas and finances involved had to be protected from the manifold competition that tried, without investing in developmental costs itself, to utilize others' ideas in its own business, thus doing much harm to the innovative firms.
For our purposes, the changes in the legal realm are of particular importance. Since the end of the 18th century, various branches of commerce, such as the textile industry, were the subject of strong measures to protect models of such things as textile patterns; this was also applied to the industrial realm. In the French textile-making city of Lyon, the *Conseils de prud'hommes* were set up on March 18, 1806, whereby textile manufacturers could deposit either openly or secretly (in a closed packet) samples of their new fabrics and thus protect them. This system of deposition was soon taken up by the Rhine Confederation. After the German Reich was founded in 1871, this became effective in all of Germany, and the principle is still in effect today.
In the ensuing years, several essential legal regulations dealt for the first time with industrial, guild and technical patterns, models and inventions for the general protection of the innovator:
--1874: Trade-mark law
--January 11, 1876: Law concerning the innovator's rights to patterns and mod els.
--May 25, 1877: Patenting of inventions

Only with the last-named law did children's toys become a separate category, protected in the whole Reich by inventors' and patent laws. From then on it was possible to have innovations in tinplate toys patented as such in Germany. Regional patents, for example, which were valid only in Bavaria (as of 1825) or Württemberg (as of 1836) and thus were scarcely effective, were superseded by the new regulations.
With the law concerning the innovator's rights to patterns and models design, the object to be protected, a photograph or drawing of it, could be deposited with a local government or registry office and its protection requested. The applicant was thereby required to give the item a number from his shop, a so-called "factory number." He often applied this number to the object itself too, as a "G.M." or "Geschmacksmuster" (taste model). If the information in a firm's archives should become lost, one could still search for the factory number in the registry office. This applies not only to protected factory numbers, but also to those that are between the protected numbers.
The process of protecting a model could take some time. As a sign that he had turned in a "G.M." model, he could apply the abbreviation "dep." or "DEP" for *deponiert* (deposited) to his product to show that this item was protected or soon would be protected.[14] "G.M." numbers and "DEP" markings thus exist only since 1876.
Since May 25, 1877, if the applicant wanted to protect an invention, he could apply for a "Deutsches Reichs-Patent", abbreviated "DRP." The protection lasted for at least six years and could be extended to approximately fifteen years. After that, everyone

could use the invention. This process also took some time. Therefore the letters "DRPA" are found on some objects, to show that a German patent had been applied for, but the process had not yet been concluded.
An addition to the model and invention, protection came on October 1, 1891, when the law to protect commercial models, or *Deutsches Reichs-Gebrauchs-Muster* (abbreviated "DRGM"), came into effect. By this law, not only individual models and inventions were protected, but so were combinations of them. This protection lasted only three years, but it was very important for the tinplate toy industry. Thus "DRGM" numbers were used only since 1891. Additional markings that are sometimes found on objects are "D.R.M.R." for *Deutsches-Reichs-Muster-Rolle*, or "D.G.M.Sch." for *Deutscher Gebrauchs-Muster-Schutz*, which has the same meaning as the "DRGM" marking. Also "GES.GESCH." for *gesetzlich geschützt* (legally protected), which does not clarify whether the form, an invention, or the combination of an external appearance with an invention is supposed to be protected by law. "DRP" and "DRGM" numbers were given sequentially in the German Reich, so that on the basis of the numbers one can determine, or at least estimate, the time when the application for model or invention protection was made. For example, patent no. 5971 was issued on December 4, 1878, no. 69972 on October 23, 1892, no. 203,576 on September 13, 1907, no. 512,113 on April 15, 1930, and no. 705,809 on April 3, 1939.[15]
But these dates only indicate after what date an object could have been manufactured. A manufacturer was not obligated to produce a product immediately after the application for and granting of model protection, nor did all patented objects go into production. Other products were produced for years, even decades, and still bore the patent number, although its effect had expired after just a few years.
Another group of markings that help us answer the question of why tinplate toys were marked includes those of origin. Here too, we find that practically no early, handmade tinplate toys carry a mark of origin, so this type of identification is not traditional.
At the end of the 1880s, the British economy suffered a great crisis, with a high balance-of-trade deficit. The British authorities tried to overcome this with a "Buy British" policy. The British people were exhorted to buy only British products. To make those products recognizable, the British authorities required that all goods brought in from elsewhere be marked with "Made in..." or the like. This procedure was included in the Tariff Act of 1890.[16]
Since the German tinplate toy industry exported up to 80% of its products at that time, and Great Britain was one of its main customers, the marking of German tin toys was carried out--though very slowly. Märklin, located in Göppingen, wrote "Made in Württemberg", while Bing of Nuremberg used "Made in Bavaria." Other firms marked their products only with "Foreign" or "Foreign product." Only around the turn of the century did the "Made in Germany" marking become generally accepted. Meanwhile, the efforts of the British authorities proved to be decidedly counterproductive. Since the German firms could now be recognized as such, and since German toys had a good reputation because of their particular quality, they were sought out and bought for that very reason. The words "Made in Germany" became a sign of quality and were maintained when it was no longer required.[17] In the 1920s and 1940s the British authorities repeated their "Buy British" campaign. That is why several British toy buses were made with the "Buy British" logo, copying the real London buses.
After World War II, the exporting of German products was banned, and several countries banned the importing of German goods. But since the American toy market was extremely dependent on German imports, special regulations were introduced for German imports from the American sector. For the rebuilding of the tinplate toy industry in Nuremberg, which was in that sector, these regulations were of great importance. Without circumventing the import and export bans, the American occupation authorities allowed the sale of tinplate toys in their PX stores, which supplied the American military personnel and their families, as long as the toys came from the American occupation zone and were marked "Made in U.S. Zone Germany." This regulation applied only to

1946-47, but since the tools were still used for a long time afterward, tinplate toys with the inscription "Made in U.S. Zone Germany" were still made until well into the fifties. This marking also became a sign of quality, particularly for Nuremberg products.

Another group of markings on tinplate are the so-called "factory marks", "protective marks", "trade marks", or just "marks" for short. They could be symbols, initials, or full names that stated or suggested the name of the firm. In the German Reich there were no legal regulations that required a firm to mark its products with its name in any way. Nor were there any international requirements of this kind. But it has been found that at the end of the 19th century more and more German firms followed the trend to use identifying marks; some marks were even copyrighted by law, like those of the Nuremberg firms of Günthermann in 1892, Meier in 1894 and Falk in 1895.

This could be justified because certain firms had used their models and inventions to develop noteworthy standards of quality and excellence. This was a new phenomenon in the industrial sector. The pressure of competition between the individual firms was also tremendous. So that wholesalers, retailers and customers could quickly identify a firm from which they had already bought something, it became necessary to place a symbol of the firm on the article.

How many such marks a firm possessed, how long they used them, and whether they applied them to all of their own products or also to foreign products, was up to the firm. For example, the Nuremberg firm of Bing marketed products made by other manufacturers under its name. There are several reasons why marks were changed: a new image for the firm, product changes, changes in the firm's status (AG, OHG, GmbH), the firm's name, merging or collaboration with another firm, buying up a foreign firm's stock, etc. On the other hand, the high recognition value was a reason to keep a trade mark unchanged. There are also reasons for the simultaneous use of several trade marks: differences in quality (for example, low-priced articles sold by a well-known firm), differentiation of branches (railroads and motor vehicles), linking one group of products with a special trade mark, etc.

Many books about tinplate toys have included outlines of individual firms' histories or charts of trade marks. The latter offers the hope that, by relating a mark to a picture in the publication, an "identification" of the object in question can be made. Time spans are often provided along with these marks, indicating when the marks were used.[18] But the linking of a firm's trade mark to a particular year is not necessarily decisive. Most of the known correlations were not worked out directly, but are the result of various considerations and assumptions. For example, if a mark is shown in a firm's dated catalog, it is simply accepted that only this mark can be found on the objects shown there. But if this assumption is not correct, then one cannot date an existing object with that mark according to that catalog. In a further step, the thus "dated" object is used as a basis of comparison to identify other, unmarked objects. This method of "identification by inference" includes many imponderables, though, and becomes less definite from step to step. Yet because of the poor state of research, this process is presently the most commonly used means of gaining further information about tinplate toys.

The knowledge of most firms in regard to their own and other marks is still relatively meager. But only through comparatively thorough knowledge of the history and products of a firm can identification by means of the markings on tinplate toys become a reliable method.

It must be mentioned that there are many tinplate toys that have not been identified by the objects themselves but by their packaging, and that it was not only manufacturers who placed marks on tinplate toys, but also suppliers and wholesalers.

Footnotes:

1) There are several indications that "DEP" and "GM" markings were used synonymously by several manufacturers.
2) See Cieslik (Ed.), 1977 and 1979.
3) The term "Tariff Act" comes from the North African city of Tarifa.

The British government conducted a great portion of its trade with Africa via this port. See White, 1975, pp. 20-21.

4) The requirement of marking products for the British market naturally did not apply only to imports from Germany. “Fabriquée en France” was frequently used in France.

5) See Broder, Leinweber (Ed.), 1984, pp. 69-106; White, 1975; Pressland, 1991, pp. 210-215.

Chapter 7.
"CLASS" IN WOOD, TINPLATE AND PLASTIC
Collecting, Designing and Buying Toys Isolde M. Th. Kohl

The fascination is greater than ever for lovingly detailed model cars of painted and lithographed sheet steel, scale models of diecast zinc, action toys, and a great variety of tinplate technical toys. The very enthusiasm of the tinplate toy collector often gives the colorful pieces of metal, originally intended as toys, an almost sacred character. The relatively young age of the objects is often surprising. A good amount is being paid for toys barely thirty years old; four-figure prices are no longer rare. But not every old auto model interests collectors. The items must have grown old "with dignity." An original package or a particular detail in the model raises its charm and its price. But the real clincher is its origin from "a good family", for the criterion of collectability is above all the quality of the brand of products. The Nuremberg and Fürth toy firms have guaranteed this for many years. Whether it was the products of the legendary Bing Brothers or the famous sheet-metal toys franked with the unchanging "Schuco label" such as the "Wendeauto", Schuco"s technical masterpiece of the thirties, or the "Studioauto", based on the renowned "Silver Arrow" of Mercedes-Benz. All of these tinplate speedsters have long since taken on an aura of the "classic", and the series of exquisitely formed miniature autos was enriched by so many high points from the production of other manufacturers.

The wealth of "classics" in many areas of the designed world around us cannot be missed today. In what ways do toys stand out for which the title "classic" appears to be justified? Assembling and comparing collectors' items made of tinplate makes an initial answer obvious. In the early sheet-metal toys in particular, we see models that were an attempt to try to reproduce the real item accurately. These miniature objects were also heralds of a technicalization process, which used them to find its way into the children's playroom. A contemporary observer of Franconian toy production before World War I tellingly commented: "A firm cannot make do with the steamships that come here, but copies pictures of our warships from the Imperial shipyards in Kiel. The industrialists and distributors of Nuremberg generally take a political standpoint that is not in agreement with the expansion of the fleet, but they simply cannot ignore such a subject, even if they promote the so-called fleet movement in the process."[19]

But it is not just the transference of technical achievements into tinplate toys that excites the collector. The sheet-metal figures, especially those made since the thirties and usually spring-operated, that nod, wobble, hop, spin or simply walk are likewise popular. These mechanical tinplate toys with built-in hearts of gears and spring steel are also collectors' items with increasing values. There are animal figures, including "humanized" ones, that represent a sort of three-dimensional form of animal caricature. The comic stars of Walt Disney form a genre of their own, of course,

Chess game, 1988, designed by Susanne Kummer, a graduate of the Giebichenstein school.

but early versions of them are quite rare. Still quite young are the various abstractly formed animal figures that first appeared in the Sixties. Manufacturers such as Ernst Paul Lehmann, Georg Adam Mangold, and even the giant firm of Schuco produced the performers of this tinplate orchestra, and even increased their variety by covering their sheet-metal bodies with plush, velvet or felt. The biggest star was the Schuco Pick-Pick bird--over twenty million, in various colors, were produced between 1928 and 1970. Now, as then, this toy makes an impression with its skillfully conceived design and a clever, rather understated abstraction of form, which is precisely fitted to the qualities of the material. Perhaps that is its charm. One of the well-known experts in this area, the late Swiss collector Peter Ottenheimer, said of it: "My interest in toy cars had its origins in my interest for real cars, but it is the difference between the toy and the real thing that fascinates me so much. The life-size real thing leaves me considerably colder than the toy, which appears to be much more capable than the real car of expressing the character of a definite time and its creation. This refers very particularly to the art of printing."[20]

But may the conclusion be drawn that handcrafting competence and perfection in corresponding formative abstraction and imagination actually produce "class"? From the collector's perspective, this is probably true. After the era of the tinplate toy had ended in the seventies, the lovely relics of the "age of tin" became limitable in their extent and easy to compare and evaluate in their qualities as models. The main problem for the collector is actually whether or not he can hunt them down. Nor must the buyers go without sheet-metal toys, for many replicas of well-known models, made of metal (sometimes even with the old tools), have appeared recently. Collectors' publications portray the tinplate cars awakened to new life. The Schuco Wende-Limousine has been produced by the TRIX-Mangold firm since 1992, likewise the Kommando-Auto, the tooting Akustico and the models of the fifties, which all children of the baby-boomer generation remember as if it were only yesterday. But the diecast models in 1/43 scale are also experiencing a revival. The firm of Märklin is returning to its traditions in that realm with models of racing cars. The old British "Dinky Toys", now in the "Matchbox" classic realm, can likewise be had again as independent series for collectors and fans. In addition, there are plenty of models of fine modern cars that are produced purely as collectors' models and try to imitate the real cars. Among these offerings, we cannot overlook the numerous inexpensive imports, some of them very badly made.

Despite all this activity, tinplate and diecast toys will never be able to win back the positions that they held from the turn of the century to about thirty years ago. The materials used in toy production have changed, and thus so have the products or their designs. Plastics have become the prevailing basic material for the toy industry; toys, kits and models, made only of plastics, make up more than one third of the entire market for toys. Wooden toys are represented by some 5% of the market. Good or bad designing, in the end, is completely independent of the material. What makes the difference is exclusively the harmony of function, material and appearance. Here too, in the realms of wood and plastic, collectors' items are already being identified as "new" classics.

Bauhaus Building Game no, 9412, designed by Alma Siedhoff-Buscher, Copyright naef AG, CH-4314 Zeiningen, 1977.

Among wooden toys, this applies particularly to those that are made with bright coloring on simple geometrical forms, and they vary. Cubes, triangles and spheres provide a richness of plastic variety. Artists have created outstanding examples of good toys made of wood. These include toy designs from the Bauhaus. This school was founded by Walter Gropius in 1919 and had workshops for formative handicraft, architecture and visual art. It has provided decisive influences in design, not only in architecture or interior design, but also for objects in everyday use. A striking example of this is the Bauhaus chess set by Josef Hartwig, which is marketed today as an "Original Bauhaus Model" by the Swiss firm of naef. The shapes, with their masterful designs, reveal the functions of the playing pieces.
Every bit as fascinating in its simplicity and creative extent is the Bauhaus building game of Alma Siedhoff-Buscher (likewise now available on the market again). There are many other examples of good design in wood: superbly made hands-on toys, building blocks, railroads, animals for small children, and many fine wooden games for young people and adults. A special sort of position is held by those play objects developed by the firm of naef that, thanks to their high quality, have particularly enriched the realm of wood in relation to modern toy design. Among these objects, developed according to pedagogical criteria, the design is at least equal to the play value. These are very high-value products that attract certain age groups--in esthetic as well as play value.

Bauhaus chess set and chessboard no. 9650-9651, designed by Josef Hartwig, Copyright naef AG, CH-4314 Zeiningen, 1989.

Meanwhile, plastics has established itself firmly in the playroom. One of the classics in plastics is the LEGO building block. It stands for more than a trade name, forming a genre of its own. The nubbed basic block in uncompromising color has not been influenced by numerous variations, which are nothing more than results of market demands. The colorful plastic blocks prove to be creative not only in the playroom. LEGO blocks have already proved their advantages in other realms, such as the competition of "Automobility 2000-2010." Young designers were challenged to turn LEGO blocks into three-dimensional auto designs for the next century. The brilliance of the results speaks not only for the designers, but also for the spectrum of variations possible with plastics blocks.
In 1989 Playmobil created "Spielwelt 1900". Here the "good old days" of bourgeois life around the turn of the century were recalled, and pleasant family life was lovingly depicted in great detail. Not only was private life transformed into a proved system concept with many small utensils and a nostalgic kitchen stove, but the outside world was included too, with Pickelhaube helmets and horse-drawn carriages. But this historic creation of a better past did not find the hoped-for response in the playroom. The concept was transformed into the "classic" doll house.
The demands made on designing in the world of toys are obviously greater than ever. And despite a series of undoubtedly very well-designed toys, there are still deficits, as always. They appear to be chiefly in the still-undervalued designing responsibility for the toy as an industrial product. Toy design often appears as a by-product by involved educators, graphic designers, interior architects, or as a borderline area of the industrial designer. But there are also extremely noteworthy initiatives in this struggle for good designs. One example of this is the Hochschule für Kunst

"LOBSTAR", designed by David Adam, Stefan Schöning, Felipe Rod Clefas, Erwin Mortier, Steven Bolle, Johan van de Putte, Edelhart van Acker, 1990. Photo: LEGO GmbH

und Design, at Giebichenstein Castle in Halle on the Saale. It can grant toy designing a special position, since this is regarded there as an independent course of study--a rarity in training for industrial design. Instruction follows a totality-oriented concept in the tradition of the Bauhaus and the Ulmer Hochschule für Gestaltung. The Hochschule at Giebichenstein Castle, which unites free art, artistic handicrafts and design under one roof, is probably unique in Germany at this time. In 1915 the training facility was developed from an artisans' school through the initiative of the architect and Werkbund member Paul Thiersch. After 1945 the school returned to the traditions of the twenties and thirties--after a checkered history. The subject of industrial design has been taught there since the end of the fifties. This institute has had a model character even in DDR days. It was able to preserve this, and has preserved it to this day. The reason for its success was and is the fact that free art, artistic handicraft and design are taught along with each other. The communication between designers and creative artists, who would remain on separate courses elsewhere for the sake of purely goal-oriented training, is guaranteed from the start at Giebichenstein. The course in "play means design" has existed since the mid-sixties; even after the political changes of 1989-90, it was maintained. The program of study includes both the design of playthings--from children to seniors--and the designing of play in the environment (the formation of playing places and playing actions) as well as toys for the Reha realm, and extends to the equipping of puppet theaters and films.

An inherent ponderousness is often attributed to "DDR design," yet in the realm of designing there are astonishing alternatives. Models that were presumably conceived and developed far away from the reality of everyday production remained rarities, of course. Yet they give evidence of competence in terms of formation, function and ease of production. All in all, inspirations that cannot be overlooked in the realm of toy design have come from the traditional toymaking centers in the Thuringian Forest and the Erzgebirge in the most recent times. A further example of this is the cooperation of leading manufacturers with design students from the Ingenieurschule für Spielzeugformgestaltung in Sonneberg. Industrial designers trained there won numerous prizes just last year at a competition of the BIG Toy Factory for the development of new variations in the BIG building-block system.

The discussion of the significance of design necessarily goes far beyond its "classic" qualities. An environment that grows ever more complex is always opposed by higher demands in terms of quality. This applies not only to the material and workmanship, but also to the adequate and timely formation. The toy thus is more than ever before on the battleground between the producer, the buyer and the playing child.[21] "Design" is a very demanding concept in this respect. It has a somewhat different meaning for each of these groups, and thus a different focal point of content. For the manufacturer and the seller, design is an important sales argument, along with the quality of the product, which also has to take play value and safety into consideration. It is a sign of quality and offers possibilities for product differentiation on the market. Adult buyers who acquire model cars and toys as design

Playmobil designs. The figures at left in each pair are the older versions of the construction worker, pirate captain and clown. Photo: playmobil, geobra Brandstätter GmbH + Co. KG

objects ask chiefly for accuracy of form and workmanship, and look at esthetics, precise details or convincing stylization, and definite brand names. In series and compatible systems the goal is completeness. It is not surprising that the borders between the buying and trading/collecting of historic models and the collecting/buying of high-priced reissues and new issues have become fluid. For parents, though, still other aspects are of importance: the question of the relationship between educational value, price and quality. The design is important, but for them it is only one standpoint among many. The offerings in the stores and specialty shops, which is very vast today, often makes the final decision very difficult. Buying good toys is influenced by identifying marks such as the added "red dot", which has been awarded for many years by a jury independent of trade and industry at the Arbeitsausschuss Kinderspiel + Spielzeug in Ulm. But these recommendations too cannot protect responsible parents from doubts before making their final choices.

The relationship of the children to the designed toy is hardest to evaluate. The question of the relationship between design quality and the playing child can perhaps be approached best, perhaps, by thinking in terms of "objects for playing" rather than "toys". Educationally valuable toys do not always touch every child's heart. Again and again, children desire to own design horrors that upset even the best-intentioned observer. Such toys are often not only an optical disaster, but are sometimes also dangerous because of poor workmanship and the use of unsuitable materials.

It can only be hoped that the presently designed and produced toy, whether of plastics, wood or metal, must in no way belie its historical forerunners in terms of fascination. As for the kind of "classics" of tomorrow, the experts and collectors will decide, and it is still questionable whether their enthusiasm will still survive the "age of plastics" and its followers.

Footnotes:

1) Jürgen Cieslik, 1980, p. 82.
2) Orell Füssli Verlag (ed.), 1984, p. 9.
3) See also Periodical: das spielzeug, January 1993, p. 36. The information there could only be estimated.
4) See also: Design Forum Nuremberg (ed.), 1991. p. 131

Sand-spreading apparatus for small children, made of plastic. Designed by Susanne Kummer, working model, 1986, design not put into production.

BIBLIOGRAPHY

L. C. Beck, *Die Fabrikindustrie Nurembergs, in Festschrift zur 40.* Hauptversammlung des Vereins Deutscher Ingenieure, Nuremberg 1899.

Georg Hieronimus Bestelmeier, *Systematisches Verzeichnis eines Magazins von verschiedenen Spiel-, Kunst- und andern nützlichen Sachen,* Nuremberg 1798.

B.-H. Christiansen & U. Leinweber (ed.), *Auto, Lok und Dampfmaschine, Technische Spielwaren des 19. und 20.* Jahrhunderts, Kassel 1985.

Jürgen Cieslik, *Blechspielzeug, Eisenbahnen, Autos, Schiffe und Flugzeuge,* Munich 1980.

Marianne & Jürgen Cieslik, *Blechspielzeug-Patente,* 2 vol.: Vol. 1, 1878-1915, Bad Honnef 1977; Vol. 2, 1916-1940, Jülich-Koslar 1979.

Leo Claretie, *Les Jouets,* Paris, no date (1894).

Design Forum Nuremberg (ed.), *Dokumentation zum 5.* Nuremberger Design Gespräch, 1991.

Johann Gabriel Doppelmayr, *Historische Nachricht von den Nurembergischen Mathematicis und Künstlern,* Nuremberg 1730.

Georg Estelmann, *Die Nuremberger Spielwarenherstellung von den Ältesten Zeiten bis zur Gegenwart.* Ein Beitrag zur Wirtschaftsgeschichte Nurembergs, Diss. Frankfurt/Main 1923.

Hans Hausherr, *Der Zollverein und die Industrialisierung, in: Karl Erich Born, Moderne Wirtschaftsgeschichte,* Cologne & Berlin 1966.

Rudger Huber, *Blechspielzeug-Autos,* Motorräder, Munich 1982.

Rudger Huber, *Schuco,* Augsburg 1991.

Wolfram Metzger, *Autos, Schiffe, Zeppeline, Eisenbahn und Dampfmaschine, Die Welt im Spielzeug,* Karlsruhe-Gaggenau 1985.

Orell Füssli Verlag (ed.), *Spielzeugautos 1890-1939,* Peter Ottenheimer Collection, Zürich, Schwäbisch Hall 1984.

Ernst Plank, *Die Nuremberg-Fürther Metallspielwaren-Industrie mit besonderer Berücksichtigung der optischen Metallspielwarenindustrie,* Diss. Erlangen 1922.

David Pressland, *The Book of Penny Toys,* London 1991.

Uwe Reher, *Eisenbahnspielzeug,* Villingen, Switzerland 1977.

Karl Rosenhaupt, *DieNuremberger-Fürther Metallspielwarenindustrie in geschichter und sozialpolitischer Bedeutung,* Stuttgart & Berlin 1907.

Otto Senst, *Die Metallspielwarenindustrie und der Spielwarenhandel von Nuremberg und Fürth,* Diss. Erlangen 1901.

Christoph Stölzl, *Die Welt aus Blech,* Munich City Museum Show Catalog, Mainz 1981.

Wilhelm Heinrich Wackenroder, *Dichtung, Schriften,* Briefe, East Berlin 1984.

Elmar Wadle, *Der Rechtschutz für Muster und Musterbücher im 19. Jahrhundert--unter besonderer Berücksichtigung von Rheinland und Westfalen.* In: Ottfried Dascher (ed.), "Mein Feld ist meine Welt"--Pattern Books and Catalogs, 1784-1914, Dortmund 1984.

Ernst Wenning, *Die Nuremberg-Fürther und Bayrische Spielwarenindustrie.* In: Deutsche Spielwaren-Zeitung, December 1930.

Georg Wenzel, *Die Geschichte der Nuremberger Spielzeugindustrie,* Diss. Erlangen-Nuremberg 1967.

Georg Wenzel, *Zur Geschichte Nurembergs als Spielzeugstadt.* In: Frankenland, Zeitschrift für fränkische Landeskunde und Kulturpflege, Vol. 8, October 1991.

Gwen White, *Toys-Dolls-Automata.* Marks and Labels, London & Sydney 1975.

Periodical

"Das Spielzeug", January 1993.

The black and white photos were taken, with kind permission, from the catalogs at the Nuremberg Toy Museum.

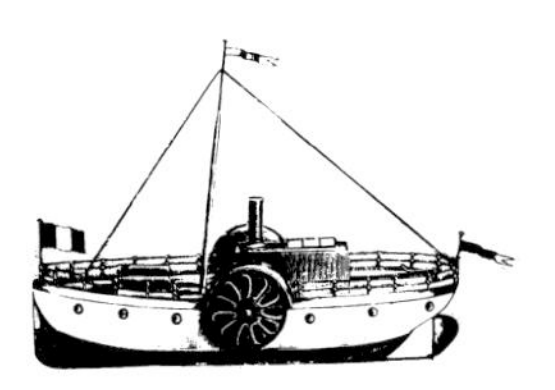

PRICE GUIDE

Values vary immensely according to the condition of the piece, the location of the market, and the overall quality of the design and manufacture. Condition is always of paramount importance in assigning a value. Prices in the Midwest differ from those in the West or East, and those at specialty antique shows will vary from those at general shows. And, of course, being at the right place at the right time can make all the difference.

All these factors make it impossible to create an absolutely accurate price list, but we can offer a guide. The prices reflect what one could realistically expect to pay at retail or auction.

The left hand number is the page number. The letters following it indicate the position of the photograph on the page: **T=top, L=left, R=right, TL=top left, TR=top right, CL=center left, CR=center right, B=bottom, BL=bottom left, BR=bottom right.** Sequential numbers following immediately after these letters indicate the position of the piece in a series of pieces reading from left to right or top to bottom. The right hand column of numbers are the estimated price ranges in United States dollars.

PAGE	POSITION	$ VALUE
2	C	75-100
15	C	500-1000
16	T & B	75-1000
17	C	100-150
18	C	400-500
19	C	300-400
20	C	50-100
21	C	50-100
22	C	150-200
23	TL	50-100
	TR	50-100
	BL	50-100
	BR	300-400
24	T	200-300 ea.
	B	50-100 ea.
25	T	100-150
	C	100-150
	B	100-150
26	C	50-100
27	C	250-300
28	T	50-100
	C	50-100
	B	50-100
29	C	50-100
30	TL	350-400
	TR	400-500
	BL	500-600
	BR	350-400
31	T	100-150
	C	100-150
	B	200-300
32	C	350-400
33	C	100-150
34	C	50-100
35	TL	200-300
	TR	50-75
	CL	200-300
	CR	50-100
	BL	50-75
	BR	200-300
40	C	1500-2000
41	C	1500-2000
42	C	400-500
43	L	1500-2000
	R	1500-2000
44	C/L	450-500
	C/R	300-400
45	C	500-1000
46	T	450-600
	B	200-300
47	L	750-1000
	R	750-1000
48	T	400-500
	C	400-500
	B	400-500
49	TL	250-300
	TR	300-350
	BL	250-300
	BR	200-250
50	C	750-1000
51	T & B	300-400
52	C	400-500
53	T	400-500
	C	400-500
	B	750-1000
54	C	750-1000
55	C	1000-1500
56	T	150-200
57	TL	1500-2000

	TR	1000 ea.
	BL	300-400 ea.
	BR (station)	500
	BR (trains)	1000 ea.
58	C	200-300
59	T	100-300
	B	200-300
60	T	50-100 ea.
60	BL	400-500
	BR	350-400
61	T	500-750
	BL	200-300
	BR	200-300
62	C	200-300
63	C	2000-3000
68	T, C, & B	400-500
69	C	250-350
70	C/L	200-300
	C/R	150-200
71	T	200-300
	B	200-300
72	T	50-100
	B	100-150
73	L	500-750
	R	350-450
74	T	400-500
	B	350-400
75	T	750-1000
	B	1000-1500 (in box)
76	C	350-400
77	TL	350-400
	TR	200-300
	BL	300-400
	BR	200-300
78	L	300-400
	TR	750-1000
	BR	400-500
79	C	500-750
80	C	250-350
81	C	350-450
82	C	250-350
83	T	300-500
	B	300-400
84	C	400-500
85	C	750-1000
86	TL	400-500
	TR	750-850
	BL	400-500
	BR	400-500
87	T	50-100
	C	50-100
	B	50-100
88	C	750-1000
89	T	75-150
	C	100-150
	B	150-200
90	C	300-450
91	T	50-75 ea.
92	T	300-450
	B	750-1000
93	T & B	300-500
94	C	750-1000
95	C/L	1000-1500
	C/R	750-1000
96	T	400-500 ea.
	C	750-1000
	B	750-1000
97	C	1000-1500
98	TL	150-200
	TR	200-300
	BL	150-200
	BR	100-200
99	T	300-450
	B	300-450
100	T	250-300
	C	250-350
	B	300-400
101	T	50-100
	C	50-100
	B	50-100
102	TL	250-350
	TR	250-350
	BL	300-400
	BR	200-350
103	T	50 ea.
	B	50-100
104	C	750-1000
105	C/L	350-500
	C/R	500-600
106	C	200-300
107	TL	350-400
	TR	750-1000
	BL	400-500
	BR	400-500
108	T	50-100
	C	50-100
	B	50-100
109	T	200-300
	B	200-300
110	T	50-100
	B	50-100
111	C	50-100
112	T	50-100
	C	50-100
	B	50-100
113	C	200-300
114	TL & TR	1000-3000
	BL	400-500
	BR	750-1000
115	TL	750-1000
	TR	1000-1850
	BL	1000-1250
	BR	750-1000
116	C	750-1000
117	C	200-300